BIODIESEL SOOT

Tribology, Properties, and Formation

XIANGUO HU
Institute of Tribology, Hefei University of Technology, Hefei, PR China

CHUAN LI
School of Chemistry and Material Engineering, Chaohu University, Hefei, PR China

ELSEVIER

Elsevier
Radarweg 29, PO Box 211, 1000 AE Amsterdam, Netherlands
The Boulevard, Langford Lane, Kidlington, Oxford OX5 1GB, United Kingdom
50 Hampshire Street, 5th Floor, Cambridge, MA 02139, United States

Library of Congress Cataloging-in-Publication Data
A catalog record for this book is available from the Library of Congress

British Library Cataloguing-in-Publication Data
A catalogue record for this book is available from the British Library

ISBN: 978-0-12-820553-2

For information on all Elsevier publications
visit our website at https://www.elsevier.com/books-and-journals

Publisher: Matthew Deans
Acquisitions Editor: Dennis McGonagle
Editorial Project Manager: Rachel Pomery
Production Project Manager: Nirmala Arumugam
Cover Designer: Victoria Pearson

Typeset by SPi Global, India

Contents

Preface

Biodiesel is a renewable engine fuel considered to be an excellent substitute for petrodiesel. Engines powered by biodiesel exhibit reduced greenhouse emissions. However, the generation of soot as a byproduct of the combustion process cannot easily be avoided. Biodiesel soot can contaminate the lubricating oil system of an engine, increasing the viscosity of the lubricating oil and engine wear. The energy efficiency and service life of the engine are significantly affected by the influence of biodiesel soot on the lubrication performance of engine lubricants. Therefore, biodiesel soot should be further investigated.

This book provides state-of-the-art information on the tribology, properties, and formation of biodiesel soot. Chapter 1 provides a general and detailed overview of elementary biodiesel. Further, it presents the preparation of biodiesel using homogeneous and heterogeneous catalysts, the basic characteristics of soot and its formation, and the hazardous effects of soot. Chapter 2 presents the morphology, structure, composition, and aggregation properties of biodiesel soot, and Chapter 3 covers the effects as well as the tribological behavior and mechanisms of biodiesel soot. Chapter 4 presents the dispersion characteristics of biodiesel soot and the tribological behavior of biodiesel soot dispersion in lubricating oils. Chapter 5 covers active control of the tribological behavior of biodiesel soot, and Chapter 6 presents the functional transformation of biodiesel soot.

We would like to thank Dr. Kunhong Hu, Dr. Yufu Xu, Dr. Hengzhou Wo, Dr. Jinsi Chen, Dr. Qiongjie Wang, Dr. Xiangyang Wang, Dr. Enzhu Hu, Dr. Tianxia Liu, Dr. Xiaonan Deng, Dr. Yubin Peng, Mr. Bo Wu, Mr. Weimin Feng, Mr. Qiangqiang Zhang, Mr. Bao Li, Mr. Yong Xu, Mr. Dongrui Yu, Mr. Kai Wu, Mr. Jian Geng, Mr. Bangbang Wang, Mr. Zhongyang Li, Mr. Youwei Fang, Mr. Jingyuan Yu, Mr. Zhichao Liu, Mr. Xiangyang Li, Mr. Yan Jiao, Mr. Quan Zheng, Mr. Bin Shi, Ms. Yuchao Tang, Mr. Lei Wang, Ms. Ziyan Lu, Ms. Zhenzhen Cao, Mr. Zhiqiang Tu, and Ms. Nana Zhou for their assistance during experimental analyses and discussion. We would also like to express our warm gratitude to Dr. Rachel Pomery, Dr. Peter Adamson, and Dr. Swapna Praveen as well as the team at Elsevier for their cooperation and efforts in developing this book. Financial support was provided from the National Natural Science Foundation of China (No. 51675153) and Anhui Province University Natural Science Research Project (No. KJ2019A0678).

This book will be of special interest to postgraduate students, researchers, and engineers working on biodiesel and soot particles. Furthermore, it will be helpful in ensuring the appropriate application of biodiesel in diesel engines and achieving an improved understanding of the tribological properties of biodiesel soot.

Chuan Li[a,b], Xianguo Hu[a]
[a]Institute of Tribology, School of Mechanical Engineering, Hefei University of Technology, Hefei, People's Republic of China; [b]School of Chemistry and Material Engineering, Chaohu University, Hefei, People's Republic of China

CHAPTER 1

Introduction

1.1 Elementary biodiesel

1.1.1 Basic concept

As the global economy continues to develop, the world's dependence on petroleum energy increases on a daily basis. However, oil is a nonrenewable resource and the world's supply of oil has become increasingly insufficient, leading to the development of alternative fuels [1–3]. Moreover, SO_2, NO_x, and other waste gases produced from the use of oil are primary causes of environmental pollution and the global greenhouse effect. Therefore, the development of renewable, degradable, and environmentally friendly alternatives to petroleum fuels is urgently needed. One such alternative fuel is biodiesel, which has the potential to alleviate the worldwide shortage of diesel fuel to a certain extent [4–6].

Biodiesel, in a broad sense, includes all alternative fuels produced from biomass. In the more narrow definition used in this book, biodiesel is the fatty acid low-carbon alcohol ester obtained by the transesterification reaction of renewable glyceride lipids and low-carbon alcohol [7]. Biodiesel's raw materials are extensively available; various edible animal and vegetable oils and waste cooking oil, leftovers from slaughterhouses, and even certain oil seeds and tree species can be used as the raw materials.

1.1.2 Basic properties

Biodiesel has the characteristics of renewability, easy biodegradation, and low emission of combustion pollutants and greenhouse gases. Biodiesel oil and petrochemical diesel have similar properties, although the former has many unparalleled advantages over the latter:

(1) Excellent environmental protection characteristics, including the low sulfur content of biodiesel, which results in fewer emissions of sulfur dioxide and sulfides, by ~30%. Biodiesel does not contain aromatic alkanes that cause environmental pollution; therefore, its exhaust gas is less harmful to humans compared to petrochemical diesel. Test results have demonstrated that, compared with petrochemical diesel,

Biodiesel Soot
https://doi.org/10.1016/B978-0-12-820553-2.00001-3

the use of biodiesel can reduce the incidence of cancer. Because of the high oxygen content of biodiesel, it produces less smoke when it burns, and its emissions contain ~10% less carbon monoxide than those of diesel oil smoke. Furthermore, the high biodegradability of biodiesel means it contributes less pollution to the soil and the environment in general.

(2) Improved starting performance in low-temperature engines.

(3) Good lubrication performance. When used in a fuel-injection pump, engine block, and connecting rod, the engine wear rate is smaller and the service life is longer compared to engines using petrochemical diesel.

(4) Better safety performance. Compared with petrochemical diesel, biodiesel has a higher flash point; hence it is less dangerous to transport, store, and use compared to petrochemical diesel.

(5) Good combustion performance. Its cetane number is high, which makes its combustion better than diesel.

(6) Renewable performance. Unlike oil reserves, the raw materials of biodiesel are renewable; therefore, there are no concerns regarding the sources of biodiesel being exhausted.

(7) Can be directly used for diesel engine combustion. Furthermore, diesel engines that run on biodiesel do not require any additional or special refueling equipment, storage equipment, or specialist technical training of personnel.

(8) When blended with petrochemical diesel in a certain proportion, biodiesel can reduce fuel consumption, improve power performance, and reduce tail gas pollution.

1.2 Preparation of biodiesel

1.2.1 Homogeneous catalytic method

Currently, the production of biodiesel by transesterification is primarily realized using the supercritical methanol, enzyme catalysis, homogeneous catalyst, and heterogeneous catalyst methods [8, 9]. The supercritical methanol method entails synthesizing biodiesel via the transesterification of methanol and feed oil in a supercritical condition without a catalyst. The supercritical methanol process has the advantages of high efficiency and environmental friendliness. However, the high-temperature and high-pressure reaction conditions lead to a significant increase in the production costs, which restricts the commercialization of this method [10].

Saka et al. attempted to synthesize biodiesel using a one-step reaction of supercritical methanol as the reactant and vegetable oil as the reaction medium at high temperature and pressure [11]. Madras et al. reported that there was a close relationship between the temperature and the conversion rate of the product. By examining the conversion rate of sunflower seed oil in supercritical methanol or ethanol, it was found that the conversion rate of biodiesel increased significantly with an increase in temperature [12].

The lipase catalyzed method synthesizes biodiesel by the transesterification of feedstock oil with methanol. For example, Köse et al. used cottonseed oil as the raw material (the molar ratio of alcohol to cottonseed oil was 4:1) and 30% Novozym 435 lipase and achieved a biodiesel conversion rate of 91.5% at 50°C for 7 h [13]. Chen et al. used Novozym 435-immobilized lipase to catalyze the transesterification of soybean oil with methanol [14]. Their results showed that the yield of biodiesel could be significantly improved by controlling the drop acceleration and reaction temperature of methanol. Mao et al. reported on biodiesel synthesized by lipase immobilized on magnetic Fe_3O_4 nanoparticles. When the molar ratio of methanol to tung oil was 3:1, the amount of added catalyst was 7.5 mg/mL, the reaction time was 26 h at 25°C, and the conversion rate of biodiesel was 91.2% [15]. Iso et al. catalyzed the transesterification of vegetable oil with methanol in 1,4-dioxane with kaolin-loaded lipase [16]. In their study, the conversion rate of the vegetable oil was > 90% after a few hours. Watanabe et al. used waste edible oil containing 2.5% free fatty acid as a raw material, 4 wt% *Candida antarctica* lipase, and methanol added in three equal doses. At a reaction temperature of 30°C and a reaction time of 24 h, the yield of methyl ester was 90.4% [17].

The engineering microalgae method is an extension of the biological enzyme method, which the US National Renewable Energy Laboratory has established using modern biotechnology. In general, the lipid content of naturally occurring microalgae is only 5%–20%; however, under laboratory conditions, the lipid content of engineered microalgae can be increased to > 60%. Biological enzyme methods have the advantages of low reaction temperatures and low methanol consumption, but they have the problems of long reaction times and the fact that lipase catalysts are expensive to purchase.

The homogeneous catalyst method primarily uses a strong acid or a strong base to catalyze the transesterification of raw oil with methanol to realize the synthesis of biodiesel. For example, Marchetti et al. used sulfuric acid to catalyze the reaction of a highly acidic raw oil with ethanol.

Under the conditions of catalyst addition and a reaction temperature and molar ratio of alcohol to 2.261 wt% oil of 55°C and 6.126, respectively, a conversion rate of 96% was achieved [18]. Al-Widyan et al. reported that hydrochloric acid and sulfuric acid can catalyze the transesterification of palm oil with ethanol to synthesize fatty acid ethyl esters [19]. Sulfuric acid had better catalytic performance than hydrochloric acid, and an excess of ethanol was conducive to shortening the reaction time. Qin et al. compared the catalytic activity of sodium hydroxide and potassium hydroxide for transesterification and reported that both of these strong bases could catalyze the synthesis of biodiesel; however, the catalytic activity of sodium hydroxide was less than that of potassium hydroxide [20]. Zhang et al. used potassium carbonate to catalyze the reaction of rapeseed oil and methanol to synthesize biodiesel. Under the conditions of a 333K reaction temperature, 6:1 molar ratio of methanol to oil, 1.5 wt% catalyst dosage, and a stirring rate of 600 rev/min, the biodiesel yield was $>98\%$ [21]. Freedman et al. used a methanol-to-oil ratio of 30:1, H_2SO_4 dosage of 1 wt%, 65°C reaction temperature, and a reaction time of 50 h, and achieved an oil-conversion rate of 99%. When they used butanol and ethanol rather than methanol, the temperature required for the reaction was 117°C and 78°C, and the reaction time was 3 and 18 h, respectively [22]. In all these methods, the products of the homogeneous reaction, the excessive reactant, and the catalyst were dissolved in one solution; hence, appropriate methods must be considered to effectively separate them. The common separation method is to precipitate and separate the glycerin first, and then wash and remove methanol and sulfuric acid from the upper liquid; finally, the remaining solution is distilled and dried to obtain biodiesel. However, washing with water will produce acidic wastewater that pollutes the environment. To address this, Nye et al. suggested that n-hexane be used as an extractant to extract fatty acid methyl esters from the reaction mixture system; they can be used later in other applications [23].

Alkaline catalysts are used for transesterification. Strong bases have high catalytic activity, and their reaction rates are about 4000 times greater than those of strong acids [24]. However, alkali-catalyzed transesterification is more sensitive to free fatty acids and water, where the presence of water makes oil esters prone to saponification. Free fatty acids react with alkali to generate soap and water, which both increases the consumption of alkali and causes emulsification in the system, thus increasing the subsequent operation process and the difficulty of biodiesel refining. Therefore, the content of free fatty acids is required to be < 1% (acid value of < 2), and the water content must be < 0.5% [25]. The most commonly used strong base catalysts are NaOH, KOH, and CH_3ONa. The CH_3ONa catalysts have the advantages of mild reaction conditions, short reaction times, and high efficiency; only a small amount is required, and these catalysts are easily removed by neutralization and washing with water after the reaction occurs and they have higher activities compared to other acids, alkalis, and metal catalysts. The KOH, NaOH, and metal alkoxides have similar catalytic properties. When NaOH is dissolved in a certain amount of methanol, a certain amount of sodium methoxide can be generated. This method is advantageous in terms of promoting contact between the catalyst and reactant, which increases the catalytic efficiency and reduces the reaction time. However, it is difficult to separate a homogeneous catalyst from a reaction medium, which considerably limits its application in industrial production.

1.2.2 Heterogeneous catalytic method

To solve these problems related to homogeneous catalysts, research on biodiesel catalysts has gradually developed in the direction of heterogeneous catalysts. In addition to the advantages of high catalytic activity and good selectivity, heterogeneous catalysts can be easily separated from the reaction medium for recycling and regeneration; moreover, they have a long service life, which has attracted the attention of many researchers [26, 27]. Heterogeneous catalysts are divided into solid acids and solid bases.

A solid acid donates protons in the reaction system, and its active center has a strong ability to accept electrons. Solid acid catalysis processes can solve the problems of strong corrosion of the experimental apparatus by the catalyst and the large energy required to separate a product from a homogeneous acidic solution. These processes can synchronously convert triglyceride and free fatty acid into fatty acid low-carbon alcohol esters, with good technical economy [28]. However, solid acid catalytic technology depends on the appropriate choice of the catalyst to a considerable extent; most methods are currently still in the laboratory research stage. Solid acid catalysts have common shortcomings, including low acid concentration, poor microporous performance, hydrophilic surfaces, tendency to lose catalytic activity, and high cost.

Commonly used solid acid catalysts are Lewis acids and $SO_4^{2-}/MxOy$ composite catalysts. Lewis acid catalysts are inexpensive, can be directly purchased, and have high esterification rates. For example, Liu et al. used iron sulfate to prepare biodiesel [29]. For an iron sulfate amount of 2%, a reaction temperature of 95°C, a molar ratio of alcohol and oil of 10:1, and a reaction time of 4 h, the esterification rate of fatty acids was 97.2%. Commonly mentioned $SO_4^{2-}/MxOy$ catalysts are SO_4^{2-}/TiO_2, $SO_4^{2-}/A1_2O_3$, and SO_4^{2}/Fe_2O_3, which are primarily prepared by the precipitation of hydroxides. For example, Wang et al. prepared a solid superacid SO_4^{2-}/Fe_2O_3 catalyst using the precipitation impregnation method [30]. For a raw oil dosage of 2%, a molar ratio of alcohol to oil of 12:1, a reaction temperature of 220°C, and a reaction time of 8 h, the biodiesel yield reached over 80%; moreover, the catalyst could be reused five times (40 h), and the yield of biodiesel after the fifth usage was still over 70%. Wen et al. studied the preparation of solid acid catalyst using glucose, starch, and sucrose as raw materials [31]. Their results demonstrated that the starch-based catalyst had the best esterification effect, reaching a biodiesel yield of > 93%.

Kansedo et al. studied the catalytic performance of SO_4^{2-}/ZrO_2 in the transesterification of palm oil, mango oil, and methanol, finding conversion percentages of 82.8% and 94.1% for palm oil and mango oil, respectively [32]. Li et al. prepared a $SO_4^{2-}/ZrO_2-TiO_2/La^{3+}$ solid acid catalyst, which was used to catalyze the esterification of fatty acids with methanol. Under the conditions of La loading, for a sulfuric acid impregnation concentration, baking temperature, and baking time of 0.1 wt%, 0.5 mo/L, 550°C, and 3 h, respectively, the catalyst demonstrated high activity and the esterification yield was up to 95% [33]. Chen et al. reported that two types of solid acid catalysts, TiO_2-SO_4^{2-} and ZrO_2_ SO_4^{2-}, were used to catalyze the transesterification of cottonseed oil with methanol, respectively [34]. Their results showed that these two types of solid acids were suitably able to catalyze the conversion of a raw oil with a high acid value into biodiesel, and the catalytic activity of the solid acid catalysts was closely related to the strength of the B acid center. Wang et al. reported on a carbon-based solid acid catalyst prepared from starch and p-toluenesulfonic acid, where oleic acid (i.e., a biodiesel simulant with a high acid value) was esterified to reduce the acid content [35]. Their results demonstrated that the catalyst had good catalytic activity for the esterification of high-acid-value biodiesel. In general, these studies have demonstrated that solid acid catalysts have strong catalytic activity and can be used to synthesize biodiesel from raw oil with high acid values [36].

A solid base donates electrons to reactants because its active center has a very strong ability to supply electrons [37]. An X-type molecular sieve is a material that contains pores of near-uniform sizes; the application of molecular sieves has been extensively studied. For example, Martínez et al.

prepared a nano-CaO/NaX catalyst for preparing biodiesel from sunflower seed oil. Under the conditions of a nano-CaO load of 16%, a reaction temperature of 60°C, a molar ratio of alcohol to oil of 6:1, and a reaction time of 6 h, the conversion rate of biodiesel was 93.5% [38]. Xie et al. studied the effects of the preparation conditions of a KOH/NaX catalyst and the transesterification reaction conditions of soybean oil conversion and investigated the reusability of the KOH/NaX catalyst [39]. Their results demonstrated that for a drying temperature of 120°C and 10% loading, the conversion rate of soybean oil was 85.6% at 65°C, a 10:1 molar ratio of alcohol to oil, 3% catalyst dosage, and an 8-h reaction time. At the end of the reaction, 48.7% of the KOH/NaX catalyst was recovered by washing with cyclohexane and drying at 125°C for 2 h; this low recovery rate indicates that some of the active components of the KOH/NaX catalyst were lost. However, the conversion of the recovered catalyst reached 84.3%. Therefore, the KOH/NaX catalyst, which underwent a loss of active components, could be treated by impregnation. This allows its catalytic activity to be recovered, hence improving its reusability. Zou et al. investigated the effect of ultrasonic radiation and mechanical agitation on the activity of a KOH/NaX catalyst [40]. Their results demonstrated that under the conditions of an ultrasonic power of 90 W, a frequency of 40 Hz, a catalyst dosage of 2%, a molar ratio of alcohol to oil of 8:1, a reaction temperature of 32°C, and a reaction time of 1 h, the conversion reached 95.8%, and the reaction rate under ultrasonic radiation was 3.79 times higher than that achieved under mechanical agitation. In their experiment, ultrasonic radiation improved the effect of the KOH/NaX catalytic transesterification.

Zhang et al. prepared a NaF/CaO catalyst with CaO as a carrier, using the equal volume impregnation method, to catalyze the transesterification of soybean oil with methanol [41]. Their results demonstrated that the yield of biodiesel was 95% at 500°C, a 4-h reaction time, a 6:1 ratio of NaF to CaO, 70°C, 8% catalyst, a 9:1 ratio of alcohol to oil, and a 2-h reaction time. Qian et al. used KF/CaO in soybean oil transesterification to prepare biodiesel [42]. Their results showed that the mass fraction of fatty acid methyl esters in the product reached 99.6% under the conditions of an ultrasonic frequency of 20 kHz, an ultrasonic intensity of 1.01 W/cm^2, a molar ratio of alcohol to oil of 12:1, a catalyst dosage of 3%, a reaction temperature of 65°C, and a reaction time of 1 h. Macleod et al. impregnated a CaO carrier into solutions of $LiNO_3$, $NaNO_3$, and KNO_3, which were then dried at 110°C for 5 h and calcined at 600°C for 5 h to yield three kinds of solid base catalysts: $LiNO_3$/CaO, $NaNO_3$/CaO, and KNO_3/CaO [43]. Under the conditions of a catalyst dosage of 5%, a molar ratio of alcohol to oil of 6:1, a reaction time of 3 h, and a reaction temperature of 60°C, the conversion rate was > 99%.

Jiang et al. prepared a supported solid base of Na_3PO_4/MgO using MgO as the carrier, which was used to catalyze the transesterification of cottonseed oil to produce biodiesel [44]. Their results demonstrated that the conversion rate reached 95.3% at a blending temperature of 70°C, a baking temperature of 600°C, a baking time of 3 h, 32% loading, a reaction time of 2.5 h, a molar ratio of alcohol to oil of 15:1, a 5% catalyst dosage, and a reaction temperature of 70°C. Xing et al. prepared solid base Li_2O/MgO using the impregnation method, which was used to catalyze the synthesis of biodiesel from Pistacia seed oil [45]. Their results demonstrated that the content of methyl ester in the product reached 98% and Li_2O/MgO had good stability under the conditions of a 3% Li content in the impregnating solution, a 600°C baking temperature, a 6-h baking time, a 3-h reaction time, a 9:1 molar ratio of alcohol to oil, a 2% catalyst dosage, and a 65°C reaction temperature. After 10 repeated uses, the content of methyl ester in the product remained at > 90%. Guo et al. prepared a CaO/MgO-supported solid base catalyst from $Ca(Ac)_2$ and MgO, which was used in the transesterification of rapeseed oil and methanol. The optimum reaction conditions were a reaction temperature of 70°C, a reaction time of 4 h, a molar ratio of alcohol to oil of 6:1, and a catalyst dosage of 3%, which yielded a transesterification conversion rate of > 98% [46]. Wan et al. prepared a KF/MgO solid base catalyst via the impregnation method, which was used to catalyze the preparation of biodiesel from rapeseed oil [47]. Their results demonstrated that the conversion rate reached 79.82% under the conditions of 30% loading, a 500°C calcination temperature, a 5-h reaction time, a 15:1 molar ratio of alcohol to oil, and a 3% catalyst dosage.

Because of the defects of single metal oxide carriers, such as the weak acidity of Al_2O_3 and the low specific surface area of alkaline earth metal oxides such as MgO and CaO, researchers have begun to study composite metal oxide carriers to overcome the limitations of single metal oxide carriers. Research began only recently on the catalytic synthesis of biodiesel with solid bases supported by composite metal oxides. Currently, the main research objects are hydrotalcite composite oxides, which have the advantages of strong alkalinity, large specific surface area, and high thermal stability [48]. This suggests they may be ideal catalyst carriers. Note that Zhang et al. used Mg-Al hydrotalcite to load a KAc-impregnated solution, which was dried and calcined to produce a supported solid-based catalyst that was used to catalyze the transesterification of soybean oil with methanol [49]. Their results demonstrated that the yield of biodiesel was >80% after a 4-h reaction with KAc loading of >20%. Xu et al. prepared a KF/Mg-Al-La-supported solid base catalyst with 30% loading, which was used to catalyze soybean oil to produce biodiesel [50]. They achieved a conversion rate of 83.39%. Wang et al. prepared a Mg-Al hydrotalcite solid base catalyst via the coprecipitation method using K_2CO_3 as the support. In their experiment to prepare biodiesel from rapeseed oil, the optimum technological conditions were as follows: a reaction temperature of 60°C, a molar ratio of alcohol to oil of 12:1, a catalyst dosage of 2%, and a reaction time of 1.5 h, which yielded an oil conversion rate of 99% [51]. Trakarnpruk et al. used Mg-Al hydrotalcite as a carrier to support metal potassium, which was used to catalyze the transesterification of palm oil with methanol to produce biodiesel [52]. Their results demonstrated that the yield of biodiesel was 96.9% under the conditions of a 30:1 molar ratio of alcohol to oil, a 7% catalyst dosage, 6 h of reaction time, and a 100°C reaction temperature.

Liu et al. prepared a $K_2CO_3/\gamma\text{-}Al_2O_3$ catalyst, which was used to catalyze the synthesis of biodiesel from rapeseed oil and methanol. The conversion rate of the biodiesel was 98.62% under the conditions of catalyst addition, reaction temperature, molar ratio of alcohol to oil, and reaction time of 4.0 wt%, 50°C, 15:1, and 3 h, respectively [53]. Furthermore, the presence of water affects the activity of the catalyst. Trakarnpruk et al. prepared a solid alkali catalyst with Mg-Al hydrotalcite as the carrier supported by metal potassium, which was used to catalyze the transesterification of palm oil and methanol to biodiesel [54]. The yield of biodiesel was 96.9% for a molar ratio of alcohol to oil, catalyst amount, reaction time, and reaction temperature of 30:1, 7 wt%, 6 h, and 100°C, respectively. Jiang et al. prepared a supported solid base catalyst using the impregnation method with hydrotalcite as the carrier and K_2CO_3 as the precursor under the conditions of K_2CO_3 loading, blending temperature, calcination temperature, and calcination time of 20 wt%, 80°C, 600°C, and 6 h, respectively [55]. Their results showed that the catalyst demonstrated good catalytic activity for catalyzing rapeseed oil to biodiesel. Sun et al. studied the performance of a Mg-Al composite solid alkaline catalyst in the transesterification process with rapeseed oil via response surface methodology (RSM). The yield of biodiesel reached 90% after 8 h of reaction at 70°C [56].

As research in this area has developed, researchers have reported that, although solid base catalysts have good catalytic activity, their costs are high, limiting their use in actual industrial production. To solve this problem, researchers have focused on new catalyst preparation methods using biomass resources. Chakraborty et al. prepared a β-$Ca_3(PO_4)_2$ solid base catalyst from fish scales via high-temperature roasting above 900°C, and they achieved a conversion rate of biodiesel of 97.73% under optimized conditions [57]. Viriya-Empikul et al. calcined eggshells at 800°C for 2–4 h to prepare a solid base catalyst with CaO as the active component. The yield of biodiesel reached >90% [58]. Boey et al. prepared a solid base catalyst from waste clam shells, used to catalyze the synthesis of biodiesel from palm oil, resulting in a yield of biodiesel of >97% under the conditions of catalyst dosage, molar ratio of alcohol to oil, and reaction time of 4.9 wt%, 0.54:1, and 3 h, respectively [59]. Corro et al. developed a Ca-solid base catalyst using animal bones as the raw material [60]. The active component of the catalyst was $Ca_5(PO4)_3(OH)$. Furthermore, under optimized conditions, the catalyst showed good catalytic activity and reusability for waste cooking oil.

Collectively, these studies have demonstrated that biomass resources have good potential for use as raw materials for biodiesel catalysts.

1.2.3 Preparation of biodiesel using homogeneous catalyst

1.2.3.1 Introduction

Animal and plant fats and oils are typically made of triglycerides, which are esters containing three free fatty acids and the trihydric alcohol, glycerol. In the transesterification process, the alcohol is deprotonated with a base to make it a stronger nucleophile. Commonly, ethanol or methanol is used. Ethanol-derived biodiesel is produced from ethanolysis of a lipid feedstock [61]. As a reactant in biodiesel production, ethanol has many advantages over methanol: it is renewable, has low toxicity, is biodegradable, has improved lubricity, has a higher flash point, and has less viscosity, making ethyl biodiesel truly a renewable energy source [62]. Soybean oil, palm oil, rapeseed oil, cottonseed oil, tallow, and waste oil can be used to prepare biodiesel. The kinetics of transesterification of soybean oil, palm oil, and rapeseed oil were investigated earlier [63–66], and the kinetics of transesterification of cottonseed oil with methanol was also studied [52, 67]. It was found that the reaction kinetics presented some differences because of different alcohols [68].

Cottonseed oil is a renewable lipid feedstock, which can be used as the raw material to prepare biodiesel. The methods used to synthesize biodiesel include transesterification, lipase enzyme, and supercritical reactions. The most common alcohols for biodiesel production are methanol and ethanol. According to Nabi [69], a biodiesel yield of 77% was obtained after 8 h at the methanol/oil ratio of 20 wt%, temperature of 55°C, and 0.50 wt% NaOH.

The conversion of cottonseed oil at 97% (max) was observed after 24 h at 50°C with a reaction mixture containing 32.5 wt% t-butanol (solvent), 13.5 wt% methanol, 54 wt% oil, and 0.017 g enzyme (g/oil) [70]. Methyl ester 97% was achieved with a reaction of 14 h, temperature of 40°C, and methanol/oil ratios of 14:1 by molar and 30 wt% enzyme in mixture [71]. Several articles [72–75] have addressed transesterification of cottonseed oil with methanol. Several studies [76–78] developed the kinetics of transesterification of cottonseed oil and methanol or ethanol with various reaction conditions. The properties and performance of biodiesel produced by cottonseed oil and methanol were investigated in two papers [79, 80].

In previous investigations, the alcohol was commonly not ethanol but methanol, for two reasons: (1) the price of methanol is lower than that of ethanol; (2) the methanolysis reaction is more rapid than the ethanolysis reaction. Methanol is generally derived from coal or natural gas; however, ethanol is fermented from reproducible agricultural resources such as rice or sugar cane, making it a renewable fuel. If ethanol were widely used to prepare biodiesel, the reliance on petroleum-based alcohols could be reduced. Therefore, biodiesel produced from ethanol is a sustainable fuel. A maximum yield of 98% was obtained after 0.5 h with potassium hydroxide of 1.07 wt%, ethanol/cottonseed oil molar ratio of 20:1, and a temperature of 25°C [81]. The conversion rate of transesterification reached 86% with ethanol/rapeseed oil molar ratio of 5:1, catalyst amount of 10 wt%, reaction temperature of 40°C, amount of n-hexane of 15 wt%, and a reaction time of 36 h [82]. As analyzed in the literature [83], the base-catalyzed formation of ethyl ester is difficult compared to the formation of methyl esters, due to the formation of a stable emulsion during ethanolysis.

At room temperature, methanol and ethanol are not miscible with triglycerides. The catalyst is generally solubilized into an alcohol layer, and the triglycerides react with alcohol in the alcohol layer. The transesterification reaction is reversible. The rate of transesterification depends on the concentration of triglycerides dissolved in the alcohol layer [84]. In order to promote the transesterification reaction process and production efficiency, a cosolvent can be introduced, with the reaction taking place in a homogeneous phase with an obviously enhanced reaction rate. The cosolvent can be classified as a matter cosolvent and an energy cosolvent. Matter cosolvent refers to the cosolvent in the form of substance, mainly various organic solvents; energy cosolvent refers to the cosolvent in the form of energy, mainly including power agitation and electromagnetic shortwave. The characteristics of cosolvents, including ethers, tertiary butanol, aromatic hydrocarbons, and alkane, have been investigated and compared with each other [85]. Tetrahydrofuran (THF) can be used as a cosolvent for transesterification because it can be easily separated from the reaction mixtures. The boiling point of THF is 66°C, which benefits the distilling separation. Employing THF allows an alkali-catalyzing homogeneous phase to take place.

The reaction kinetics of transesterification of cottonseed oil with ethanol has not been reported elsewhere. With this consideration, this section covers the kinetics of transesterification. The dynamic equations and parameters have been widely investigated. Furthermore, the study discussed here investigated alkali-catalyzed ethanolysis of cottonseed oil with ethanol in the presence of THF as cosolvent. Ethanol/cottonseed molar ratio, reaction temperature, content of catalyst, reaction time, and THF amount are presented in detail.

1.2.3.2 Experiments

Materials

For the experiments presented here, ethanol absolute, potassium hydroxide, and dimesoperiodic acid at the highest grades were provided by Hefei Chemicals Co. Cottonseed oil (acid value: 0.30 mg KOH/g, saponification value: 187 mg KOH/g, density: 925 kg/m^3, triglycerides: 95 wt%, average molecular weight: 900.1) was provided by Hefei Oil & Fats Co. All the chemicals used were analytical reagent grade.

Experimental methods

The experiments were conducted in a 500-mL flask connected to a reflux condenser and the reaction mixture was agitated by a magnetic stirrer at 300 rpm. Ethanol and KOH calculated as per experimental design were blended and then mixed with the cottonseed oil. The reaction mixture was heated to the reaction temperature when placed in an oil bath warmed at the reaction temperature. The reaction was stopped by removing the oil bath and adding 0.1 mol/L hydrochloric acid as buffer solution. The upper-layer mixture was removed and then purified. The yellow opaque liquid (i.e., biodiesel) was obtained after neutralization, sedimentation, washing, filtration, and drying.

The glycerin content of the resultant mixture is equal to the conversion rate of triglycerides [86]. The glycerin yield (Eq. 1.1) was calculated using the following equation:

$$G(\%) = RG(\%) - MG(\%) \tag{1.1}$$

where G (%) is the total yield of glycerin, RG (%) is the content of glycerin in the resultant mixture, and MG (%) is the content of glycerin in cottonseed oil.

The conversion rate of triglycerides can be calculated using the following Eq. (1.2):

$$\text{Conversion}\ (\%) = \frac{\text{MG} - \text{BG}}{\text{MG}} \times 100\% \tag{1.2}$$

where conversion (%) is the conversion rate of triglycerides; MG is the percent of glycerin ester in cottonseed oil; BG is the percent of glycerin ester in biodiesel.

1.2.3.3 Kinetic investigations of biodiesel from transesterification of cottonseed oil with ethanol

Establishment of kinetic model

Triglycerides are reacted with an alcohol, such as ethanol, to give ethyl esters of fatty acids and glycerol:

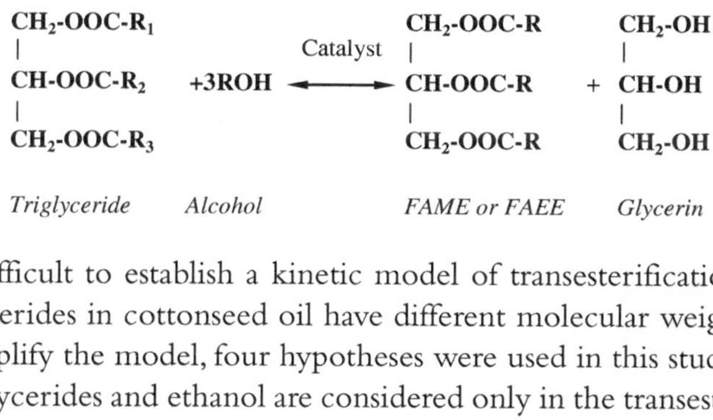

Triglyceride *Alcohol* *FAME or FAEE* *Glycerin*

It is difficult to establish a kinetic model of transesterification because the triglycerides in cottonseed oil have different molecular weights. In order to simplify the model, four hypotheses were used in this study:

(1) Triglycerides and ethanol are considered only in the transesterification reaction, ignoring other reactants and side reactions in the mixture.
(2) Isomerides of triglycerides have the same reaction rate and mechanism.
(3) There is no effect on KOH concentration from the saponification reaction.
(4) The transesterification reaction is simplified as a reversible reaction.

Based on these assumptions, the reaction rate of the transesterification of cottonseed oil with ethanol can be expressed by the following Eq. (1.3):

$$r_A = k_1 c_A^{\alpha} c_B^{\beta} \tag{1.3}$$

where r_A is the reaction rate, k_1 is the rate constant, c_A, c_B are the concentrations of triglycerides and ethanol, respectively, and α, β are the reaction order numbers.

Theoretically, the transesterification reaction requires 3 mol of ethanol and 1 mol of triglycerides to prepare 3 mol of ethyl esters of fatty acids and 1 mol of glycerol by stoichiometry [87]. Because the reaction is reversible, the excess ethanol was used to shift the equilibrium to a positive reaction and obtain a higher conversion.

According to the previous analysis, $k_1 c_B^\beta$ can be regarded as a constant k. The reaction rate can then be simplified as shown in Eq. (1.4) :

$$r_A = k c_A^\alpha \tag{1.4}$$

The conversion rate of triglycerides is expressed as x. The initial concentration of triglycerides in the mixture is defined as c_{A0}. The instantaneous concentration of triglycerides is described as c_A. Thus, c_A can be determined according to the following equations:

$$c_A = c_{A0}(1-x) \tag{1.5}$$

$$r_A = -\frac{dc_A}{dt} = \frac{d\left[c_{A0}(1-x)\right]}{dt} = c_{A0}\frac{dx}{dt} = k c_A^\alpha \tag{1.6}$$

Substitute Eq. (1.5) into Eq. (1.6):

$$\frac{dx}{dt} = \frac{k}{c_{A0}}\left[c_{A0}(1-x)\right]^\alpha = k_2\left[c_{A0}(1-x)\right]^\alpha \tag{1.7}$$

where $k_2 = k/c_{A0}$. Using the natural log function on both sides of Eq. (1.7):

$$\ln\frac{dx}{dt} = \alpha \ln\left[c_{A0}(1-x)\right] + \ln k_2 \tag{1.8}$$

Defining $x(\ln[c_{A0}(1-x)])$ − coordinate, $y\left(\ln\frac{dx}{dt}\right)$ − coordinate, Eq. (1.8) can be expressed as a linear equation by linear fitting. Constants k_2, k and order number α can be obtained by substituting experimental data into Eq. (1.8).

Kinetic parameters for transesterification

The transesterification reaction was carried out under the following conditions: the molar ratio of ethanol and cottonseed oil was 30:1, the amount of KOH catalyst was 1.25 wt% of the cottonseed oil, reaction time was 90 min, and reaction temperatures were 40, 60 and 78°C, respectively. Fig. 1.1 shows the effect of reaction temperature on conversion of triglycerides in the transesterification reaction.

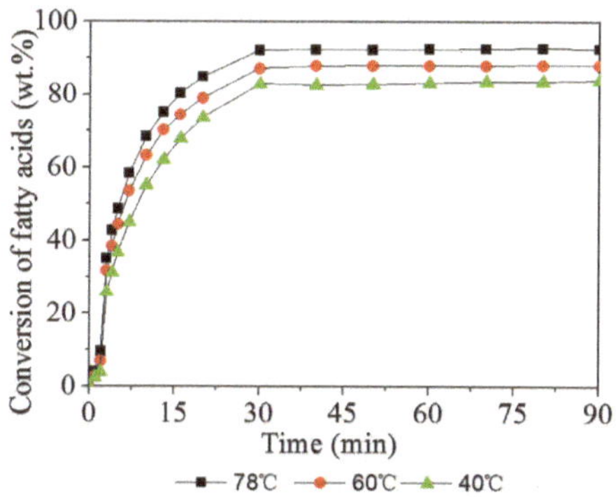

Fig. 1.1 Effect of reaction temperature on conversion of fatty acids.

It was found that the triglycerides conversion was lower in the first 3 min because the cottonseed oil and ethanol are not mutually soluble liquids. This result was also consistent with other investigations [88, 89]. A rapid increase in the conversion of triglycerides appeared within 30 min. There was no increase in conversion after 30 min. This indicated that the transesterification reaction reached equilibrium.

Using Origin 6.0 Professional, the equation was obtained for the first 30 min with temperatures at 78, 60, and 40°C, respectively. The transesterification formulas at different temperatures are shown in Table 1.1. Substituting the formulas into Eq. (1.8), α, k_2 and k can be found by using the least squares method.

Based on this method, other kinetic parameters for transesterification were obtained at 78, 60 and 40°C. The reaction rate constants and order numbers for the transesterification are displayed in Table 1.2.

Table 1.1 Transesterification formulas at different temperatures.

Temperature (°C)	Fitting equations	$\frac{dx}{dt}$
78	$x = \frac{1.1274t}{6.5263 + t}$	$\frac{(1.1274 - x)^2}{7.3578}$
60	$x = \frac{1.0802t}{7.1373 + t}$	$\frac{(1.0802 - x)^2}{7.3578}$
40	$x = \frac{1.1068t}{10.0491 + t}$	$\frac{(1.1068 - x)^2}{11.1223}$

Table 1.2 Reaction rate constant and order number for transesterification.

Temperature (°C)	k_2	K (L/mol/min)	α
40	0.1020	0.0996	1.54
60	0.1219	0.1190	1.51
78	0.1317	0.1286	1.29

The transesterification reaction of cottonseed oil with ethanol is a pseudo-second-order reaction during the first 30 min. Ethanol has almost no effect on the conversion of triglycerides because of excess alcohol. The transesterification reaction can be described as a zero-order reaction after 30 min.

Activation energy of transesterification

The Arrhenius activation energy (Eq. 1.9) can be expressed as follows:

$$k = k_0 e^{-\frac{Ea}{RT}} \tag{1.9}$$

where k is the frequency factor, Ea is the activation energy, and R is the general gas constant.

Using the natural logarithmic function on both sides of Eq. (1.9) gives:

$$\ln k = -\frac{Ea}{R} \times \frac{1}{T} + \ln k_0 \tag{1.10}$$

The linear relationship of ln k and $1/T$ is shown in Fig. 1.2. The slope of the line is equal to Ea and the numerical value of the intercept represents the frequency factor k_0.

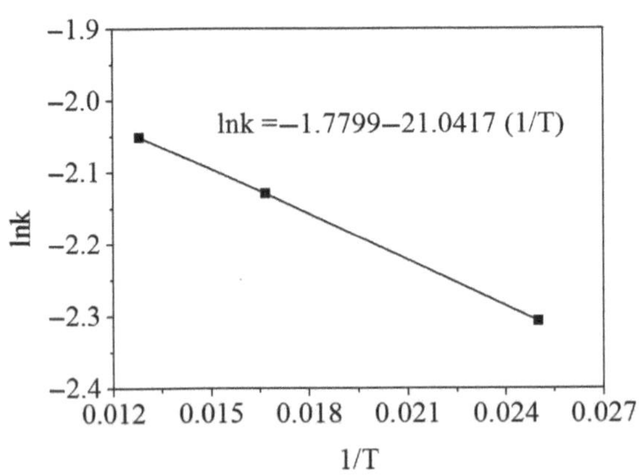

Fig. 1.2 Relation between $1/T$ and ln k.

Using linear fitting, the line in Fig. 1.2 can be expressed as follows:

$$\ln k = -\frac{22.0417}{T} + \ln k_0 \tag{1.11}$$

where $R = 0.98031$, $E_a = 21.6075$ kJ/mol, and $k_0 = 0.1687$.

Based on the previous reasoning, the equation for the kinetics of transesterification at 78°C of cottonseed oil with ethanol can be summarized as

$$-r_A = 0.1687e^{0.2826}c_A^{1.2856} \tag{1.12}$$

Generally, the activation energy of transesterification of vegetable oil with an alkaline catalyst was 33–83.17 kJ/mol [90]. The activation energy of transesterification of cottonseed oil with methanol was 10.88 kJ/mol. But the activation energy of transesterification of cottonseed oil with ethanol was 21.6075 kJ/mol. This shows that the transesterification with ethanol was more difficult to achieve than with methanol.

1.2.3.4 Biodiesel production by transesterification of cottonseed oil with ethanol using tetrahydrofuran as cosolvent

Effect of ratio of ethanol to cottonseed oil

The molar ratio of ethanol to cottonseed oil is one of the most important factors affecting the conversion of glycerin ester. Although the stoichiometric ratio requires 3:1, transesterification is commonly carried out with an extra amount of alcohol in order to shift the equilibrium to the proposed ethyl ester. The molar ratio of ethanol/cottonseed oil is 56:1 in the literature [91]. When the molar ratio of ethanol/cottonseed oil was small, the transesterification reaction was slower. However, the cost of mixture separation was high with an increased molar ratio. The ethyl ester was produced by cottonseed oil transesterification with ethanol, with reaction temperature of 60°C, KOH 1.25 wt%, reaction time of 1 h, and amount ratio of THF/cottonseed oil of 0.75:1; the molar ratio of ethanol/cottonseed oil varied from 6:1 to 16:1. Fig. 1.3 shows the effect of the molar rate of ethanol to cottonseed oil on the conversion of triglycerides.

The conversion of triglycerides is enhanced with an increase in the molar rate of ethanol to cottonseed oil, because transesterification is a reversible reaction. The high content of ethanol can shift the reaction to the proposed effect. The maximum conversion of triglycerides of 86.53% was obtained when the molar rate of ethanol to cottonseed oil was 12:1. The conversion of triglycerides declined when the molar rate of ethanol to cottonseed oil exceeded 12:1. The decrease in content of KOH was due to excess ethanol in the mixture, which brought about a decrease in the catalyst active center.

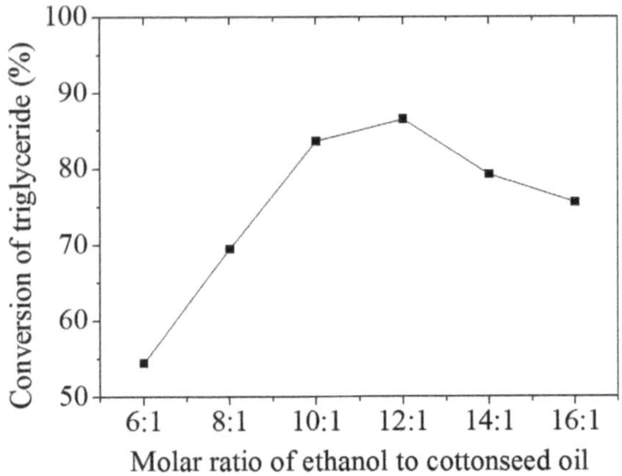

Fig. 1.3 Effect of molar ratio of ethanol to cottonseed oil on the conversion of triglycerides.

Effect of concentration of catalyst

The maximal conversion of triglycerides was achieved with a molar rate of ethanol to cottonseed oil of 12:1, reaction temperature of 60°C, reaction time of 1 h, and amount ratio of THF/cottonseed oil of 0.75:1. The optimum concentration of catalyst KOH was 1.25 wt% of cottonseed oil. The effect of KOH concentration on the conversion of triglycerides is shown in Fig. 1.4.

The concentration of KOH was explored over a range of 0.25–1.5 wt%. As KOH concentration increased from 0.25 wt% to 1.25 wt%, the

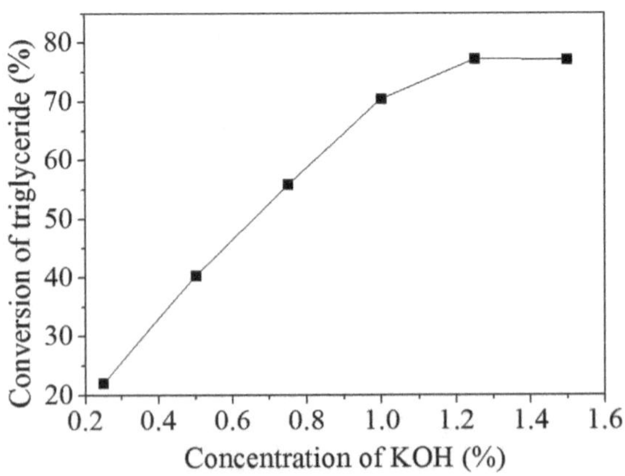

Fig. 1.4 Effect of concentration of catalyst KOH on the conversion of triglycerides.

conversion increased from 22.07% to 77.25%, respectively. However, the conversion decreased with increase in the concentration of catalyst; this is due to the excess KOH resulting in the formation of soap and increased viscosity, which hinders the course of transesterification.

Effect of temperature

Fig. 1.5 shows the effect of temperature on the conversion at an ethanol/oil ratio of 12:1, reaction time of 1 h, amount ratio of THF/cottonseed oil of 0.75:1, and a 1.25 wt% concentration of KOH.

Upon increasing the temperature from 40°C to 60°C, the conversion increased 12.18%. At a temperature of 60°C, there was a slight reduction in the conversion. Although the required time to reach a reaction balance could be lessened with high temperature, higher temperature has no benefit for increasing conversion. Furthermore, high temperature enhanced both transesterification and saponification reactions, resulting in the decrease of conversion and the difficulty of product purification.

Effect of reaction time

Fig. 1.6 shows the effect of reaction time on the conversion. The highest conversion of triglycerides was obtained after 2.5 h at the ethanol/oil ratio of 12:1, amount ratio of THF/cottonseed oil of 0.75:1, temperature of 60°C, and concentration of KOH of 1.25 wt%.

Increasing reaction time from 1 to 2.5 h increased the conversion by only 4.39%. There was no obvious effect of prolonging reaction time on conversion increase. According to Schwad [92], the transesterification

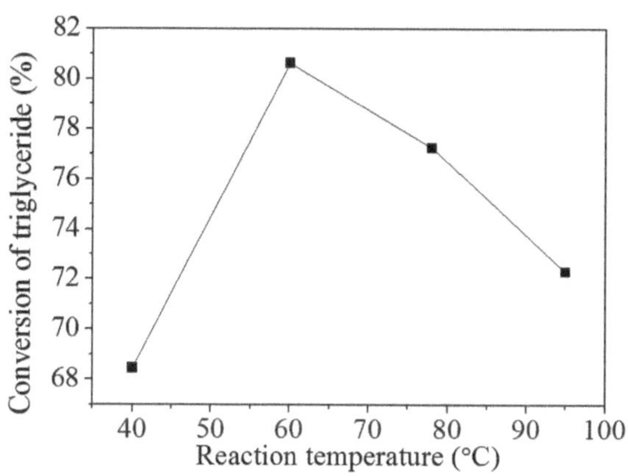

Fig. 1.5 Effect of temperature on the conversion of triglycerides.

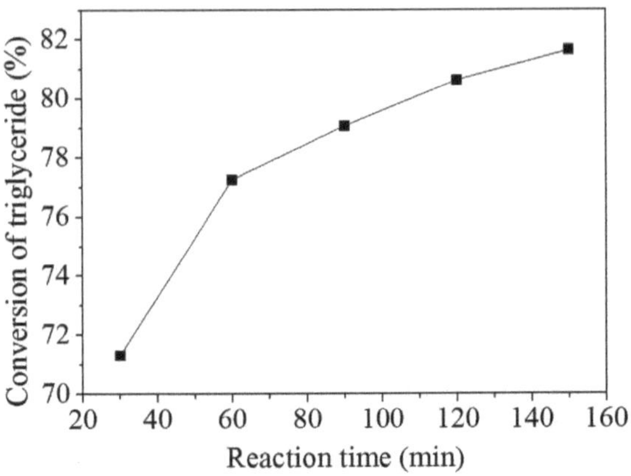

Fig. 1.6 Effect of reaction time on the conversion of triglycerides.

reaction could be completed within 30 min, and a conversion of 71.3% was obtained after 30 min. The conversion increased only 10.34% during the next 120 min.

Effect of cosolvent

Shown in Fig. 1.7, the conversion could be observed at an ethanol/oil ratio of 12:1, temperature of 60°C, reaction time of 1 h, and concentration of KOH of 1.25 wt%. The ratio of the amounts of THF/cottonseed oil was in the range of 0.25–1.25. The conversion increased rapidly with THF/cottonseed oil increasing, but the conversion decreased markedly when the THF/cottonseed oil ratio exceeded 0.75:1.

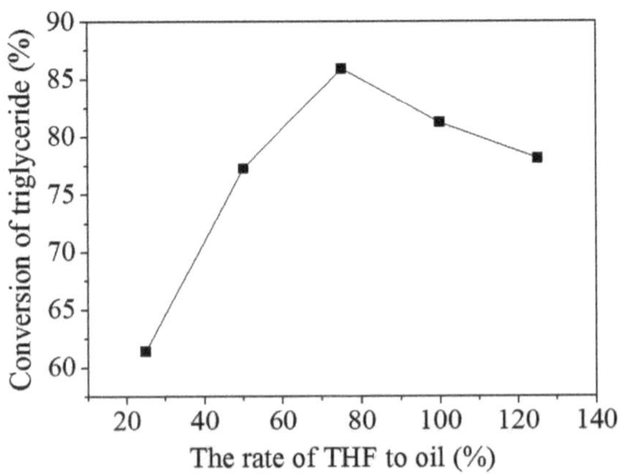

Fig. 1.7 Effect of cosolvent on the conversion of triglycerides.

The highest conversion of 85.92% was obtained at the ratio of THF/cottonseed oil of 0.75:1. The concentration of KOH was only 1.25 wt% in this mixture, but a high conversion was observed in the experiment. Test results showed that THF could accelerate the transesterification reaction and increase the conversion of triglycerides.

Optimization of processing parameters

Single factor experiments have indicated that ethanol/oil, catalyst concentration, temperature of reaction, time of reaction, and amount ratio of THF/cottonseed oil all have different effects on the conversion of triglycerides, but there is no conclusion as to which factor has the leading effect. Orthogonal test analysis methodology was chosen to optimize transesterification reaction parameters. Table 1.3 shows the five factors with three levels of each factor. Table 1.4 shows the results of the orthogonal test.

Test results show that the order-of-factors effect on conversion is as follows: ethanol/oil > reaction temperature > concentration of cosolvent > reaction time > amount of catalyst. The optimum process parameter is $A_3B_3C_2D_3E_2$. The maximal conversion of 88.73% was observed after 1.5 h with ethanol/oil of 14:1, reaction temperature of 78°C, amount of KOH of 1.25 wt%, and ratio of THF/oil of 1:1.

1.2.3.5 Conclusions

The transesterification reaction of cottonseed oil and ethanol was carried out with a molar ratio of ethanol to cottonseed oil of 30:1, amount of KOH catalyst of 1.25 wt%, and reaction temperature of 40–78°C. Reaction equilibrium occurred after 30 min. The transesterification reaction between cottonseed oil and ethanol was a pseudo-second-order reaction during the first 30 min and a zero-order reaction after 30 min. Reaction rate constants for the transesterification of cottonseed oil with ethanol at 40, 60, and 78°C were 0.0996, 0.1126, and 0.1286 L/mol/min respectively, and

Table 1.3 Five factors with three levels of every factor.

Levels	Ethanol/oil A	Concentration of cosolvent B (wt%)	Reaction temperature C (°C)	Reaction time D (min)	Amount of catalyst E (wt%)
1	10	0.5	60	30	1.0
2	12	0.75	78	60	1.25
3	14	1.0	95	90	1.5

Table 1.4 Results of orthogonal test.

Serial number	A	B	C	D	E	Conversion (%)
1	1	1	1	1	1	54.87
2	1	2	2	2	2	75.58
3	1	3	3	3	3	75.43
4	2	1	1	2	2	65.27
5	2	2	2	3	3	78.13
6	2	3	3	1	1	76.31
7	3	1	2	1	3	80.62
8	3	2	3	2	1	84.13
9	3	3	1	3	2	78.23
10	1	1	3	3	2	64.72
11	1	2	1	1	3	60.27
12	1	3	2	2	1	68.56
13	2	1	2	3	1	81.64
14	2	2	3	1	2	84.79
15	2	3	1	2	3	75.59
16	3	1	3	2	3	83.4
17	3	2	1	3	1	78.22
18	3	3	2	1	2	86.44
K1	399.43	430.52	412.45	435.08	443.73	
K2	461.73	458.12	470.97	452.53	455.03	
K3	491.04	460.56	468.78	456.37	453.44	
k1	66.58	71.75	68.74	72.51	73.96	
k2	76.96	76.35	78.50	75.42	75.84	
k3	81.84	76.76	78.13	76.06	75.57	
R	15.26	5.01	9.76	3.55	1.88	

the activation energy of transesterification was 21.6075 kJ/mol. The transesterification reaction of cottonseed oil with ethanol was more difficult to achieve compared with that of methanol. Triglycerides and ethanol could easily mutually dissolve with THF in the reaction mixture. THF is beneficial in increasing triglyceride conversion and lessening the time to reaction balance. THF can also be used repeatedly by separating it from the mixture.

According to the orthogonal test, the leading factor is ethanol/oil. The next most influential factors, in order, are reaction temperature > concentration of cosolvent > reaction time > amount of catalyst. The highest conversion of 88.73% was obtained by transesterification with ethanol/oil of 14:1, reaction time of 1.5 h, reaction temperature of 78°C, amount of KOH of 1.25 wt%, and ratio of THF/oil of 1:1. Higher conversion of triglycerides, shorter reaction time, and a smaller amount of catalyst are the main advantages of transesterification with a cosolvent.

1.2.4 Preparation of biodiesel using heterogeneous catalyst

1.2.4.1 Introduction

Biodiesel is made from animal fats or waste cooking oil by transesterification with other short-chain alcohols. It is a mixture of fatty acid methyl esters, which makes it a useful form of renewable energy. Currently, the industry mainly uses homogeneous catalysts, such as potassium hydroxide, sodium hydroxide, or a few others, to synthesize biodiesel. Using a homogeneous catalyst has several advantages, such as fast response rate and high conversion rate. However, this type of catalyst is difficult to separate from the reacting system, and the posttreatment processes produce large amounts of wastewater. To solve the problems associated with homogeneous catalysts, many researchers have developed a variety of heterogeneous catalysts, including solid acid, solid base, and immobilized biological catalysts [93]. However, these catalysts are usually expensive and are not widely used. Thus more research is needed on these catalyst materials to develop a heterogeneous catalyst with potential industrial applications.

An important way to reduce the cost of biodiesel production is to use the extensive worldwide biomass resources as catalysts. Many solid base catalysts that use biomass as the catalyst support, such as Li_2CO_3/rice husk ash [94], KF/biont shell [95], and KI/oyster shell [96], have been used. As a type of broadleaved wood ash, camphor tree ash contains high amounts of calcium. Therefore, as a catalyst support, camphor tree ash can disperse and support the active component, and also provide additional activity sites (CaO). However, the preparation and application of camphor tree ash as a catalyst support with a heterogeneous catalyst have not been reported. Hence, the purpose of this work was to evaluate camphor tree ash as a support for K_2CO_3 to prepare a solid base catalyst for biodiesel production.

1.2.4.2 Experiments

Materials

Soybean oil was obtained from China Oil and Foodstuffs Corporation. Inductively coupled plasma mass spectrometry (ICP-MS, X Series 2, Thermo Fisher Scientific) was used to analyze the concentrations of the main metal elements of camphor tree ash. The main metal element composition of camphor tree ash was 12.05% Ca, 1.82% Mg, 2.70% Al, 0.23% Na, and 1.22% K. All the chemical reagents were analytically pure.

Catalyst synthesis

The combustion of aged camphor tree leaves naturally generates camphor tree ash. The camphor tree ash was calcined at high temperature in the first calcination process for 2 h. Typically, the first-calcination camphor tree ash was added and impregnated into a K_2CO_3 solution of a certain concentration for 12 h. The mixture was dried at 110°C for 12 h, and then placed in a box electric-resistance furnace to be calcined in a second calcination for 3 h. Field emission scanning electron microscopy (FE-SEM, SU8020, Hitachi) was conducted to analyze the internal structure and size of the camphor tree ash, catalyst precursor, and K_2CO_3/camphor tree ash catalyst.

Preparation of biodiesel

Using a certain methanol/oil molar ratio, methanol, soybean oil, and a varied quantity of catalyst, the reaction vessel was placed in a water bath under stirring using a specified time and temperature. After the reaction, the catalyst was filtered in vacuum and recovered. The crude biodiesel was separated from the upper layer, and the crude glycerol was separated from the lower layer. The glycerol content was determined using the potassium periodate oxidation method [86], because the conversion of glycerol indirectly reflected the conversion of biodiesel. The formula for the biodiesel yield Eq. (1.13) is:

$$Y(\%) = \frac{m_1 \times w}{m_2} \times 100 \qquad (1.13)$$

where m_1 (g) represents the mass of crude glycerol, w (%) represents the glycerol content, m_2 (g) represents the theoretical mass of glycerol, and Y (%) represents biodiesel yield.

1.2.4.3 Catalyst Characterization and biodiesel synthesis

Characterization of catalyst

Fig. 1.8A shows an SEM image of the camphor tree ash, which has an open, irregularly porous structure. Fig. 1.8B shows an SEM image of the camphor tree ash after the first calcining at 800°C, consisting of crystal particles that are mainly small aggregated blocks, with a smooth particle surface. Fig. 1.8C shows an SEM image of the K_2CO_3/camphor tree ash catalyst (second calcining at 500°C). Unlike the result of the first calcination, the crystal particles here have mainly irregular sheet structures with clear gaps between the particles. The typical diameter of a crystal particle of the camphor tree ash after the first calcining is about 650 nm, whereas that of the K_2CO_3/camphor tree ash catalyst is about 300 nm. This indicates that the second calcination greatly improved the particle size of the catalyst particles and the particle structure.

Effect of catalyst synthesis condition on biodiesel yield

As shown in Fig. 1.9A, the biodiesel yield gradually increases with K_2CO_3 loading. However, after K_2CO_3 loading of 50 wt%, the biodiesel yield does not increase significantly with additional loading. This result indicates that the

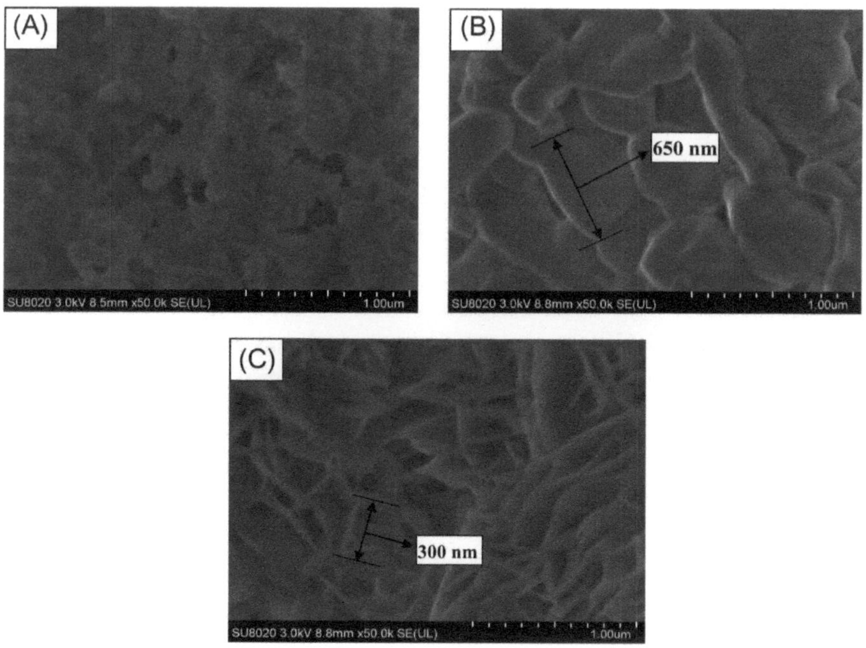

Fig. 1.8 SEM images of (A) camphor tree ash, (B) camphor tree ash after the first calcining at 800°C, and (C) K_2CO_3/camphor tree ash catalyst.

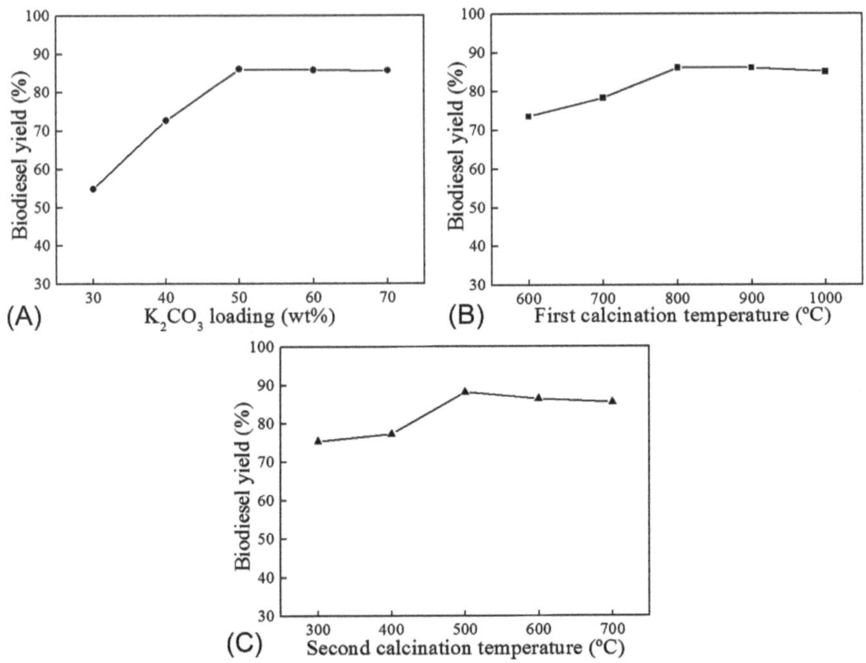

Fig. 1.9 Effect of catalyst synthesis conditions on biodiesel yield. (A) K_2CO_3 loading (first calcination temperature = 800°C, second calcination temperature = 600°C); (B) first calcination temperature (K_2CO_3 loading = 50 wt%, second calcination temperature = 600°C); and (C) second calcination temperature (K_2CO_3 loading = 50 wt%, first calcination temperature = 800°C).

catalytic center is saturated when the K_2CO_3 loading is 50 wt%. As shown in Fig. 1.9B, the biodiesel yield increases gradually with the first calcination temperature. After a first calcination temperature of 800°C, biodiesel yield decreases slightly with further increase of calcination temperature. This result indicates that the $CaCO_3$ in the camphor tree ash has been completely converted to CaO when the first calcination temperature reaches 800°C [97]. As shown in Fig. 1.9C, the second calcination temperature and the first calcination temperature changing patterns were similar. It indicates that a second calcination temperature that is too high or too low reduces the catalyst activity.

Effect of transesterification reaction condition on biodiesel yield

As shown in Fig. 1.10A, the biodiesel yield increases gradually with reaction time. After 210 min, the biodiesel yield does not increase significantly with additional reaction time, the transesterification reaction balance is gradually completed, and the forward and reverse reaction rates approach a new

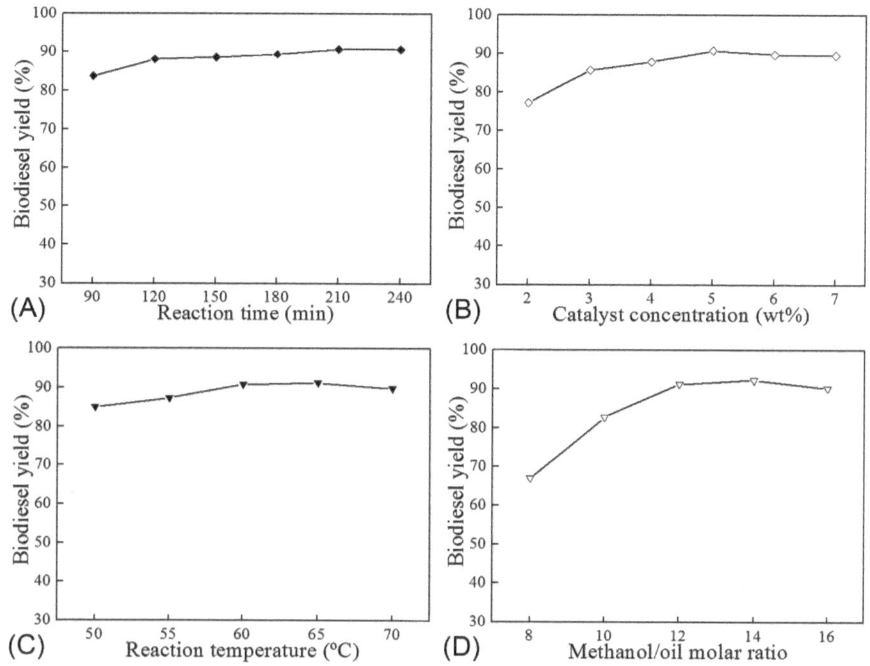

Fig. 1.10 Effect of transesterification reaction conditions on biodiesel yield. (A) reaction time (catalyst concentration = 5 wt%, reaction temperature = 60°C, methanol/oil molar ratio = 12:1); (B) catalyst concentration (reaction time = 210 min, reaction temperature = 60°C, methanol/oil molar ratio = 12:1); (C) reaction temperature (reaction time = 210 min, catalyst concentration = 5 wt%, methanol/oil molar ratio = 12:1); and (D) methanol/oil molar ratio (reaction time = 210 min, catalyst concentration = 5 wt%, reaction temperature = 65°C).

balance. As shown in Fig. 1.10B, the biodiesel yield increases gradually with increasing catalyst concentration. However, after a catalyst concentration of 5 wt%, the biodiesel yield decreases with catalyst concentration increase. This result shows that the reaction system basic active centers increase with catalyst concentration increase. When the catalyst concentration exceeds 5 wt%, the catalyst in the reaction system affects the mixing of reactants, which is not conducive to full contact between reactants and catalysts [98].

As shown in Fig. 1.10C, the biodiesel yield increases gradually with reaction temperature. After a reaction temperature of 65°C, the biodiesel yield decreases with reaction temperature increase. This result shows that, in the case of low reaction temperature, increasing the reaction temperature promotes the interdiffusion of soybean oil and methanol and increases the transesterification reaction rate. However, when the reaction temperature exceeds 65°C, the reaction system temperature exceeds the boiling point of methanol, which results in a lowering of the methanol concentration in the liquid phase [99]. As shown in Fig. 1.10D, the biodiesel yield increases gradually with increasing methanol/oil molar ratio. When the methanol/oil molar ratio reaches 14:1, the biodiesel yield reaches its maximum value (92.27%). This indicates that to improve biodiesel yield, the typical selection of an actual methanol/oil molar ratio should be far higher than the theoretical methanol/oil molar ratio of 3:1. However, when the methanol/oil molar ratio exceeds 14:1, there is too much methanol in the reaction system, which can increase the solubility of glycerin and move the transesterification reaction equilibrium toward the left [100].

Reusability of the catalyst

The K_2CO_3/camphor tree ash catalyst was prepared under optimal catalyst preparation and transesterification reaction conditions. Then, the catalyst was recovered between each reaction by filtration and thoroughly washed with methanol. After the K_2CO_3 was impregnated again with the used K_2CO_3/camphor tree ash catalyst and then calcined at 500°C for 3 h, the biodiesel yield using the treated catalyst reached 90.05%, which indicates that regeneration of the K_2CO_3/camphor tree ash catalyst can be achieved through the method of reloading the active component K_2CO_3.

1.2.4.4 Conclusions

The K_2CO_3/camphor tree ash catalyst can catalyze the transesterification reaction to synthesize biodiesel. Under optimal preparation conditions (K_2CO_3 loading, 50 wt%; first calcination temperature, 800°C; second calcination temperature, 500°C), the prepared K_2CO_3/camphor tree ash catalyst can be used for transesterification and a biodiesel yield of 92.27% can be achieved with a reaction time of 210 min, catalyst concentration of 5 wt%, reaction temperature of 65°C, and methanol/oil molar ratio of 14:1. The K_2CO_3/camphor tree ash catalyst can be regenerated by reloading the active component K_2CO_3, and the biodiesel yield can reach 90.05% after reloading.

1.3 Basic characteristics of soot and its formation

1.3.1 Brief introduction to basic characteristics of soot

1.3.1.1 What is soot?

Soot particulate matter is an important component of atmospheric fine particulate matter (e.g., PM2.5, which has particle diameters equal to or less than 2.5 μm), which is mainly derived from soot emitted by the incomplete combustion of fossil fuels and biomass [101]. Soot is one of the primary air pollutants, especially soot produced by exhaust gases emitted by transportation vehicles, which has considerable impact on the air quality of cities. Soot particles primarily contain carbon, dust, and incompletely burned hydrocarbons. The results of several studies have demonstrated that soot consists of 90% carbon, 4% oxygen, 3% hydrogen, nitrogen, sulfur, and trace metals. The size of a single soot particle is ~40 nm. However, because of the adhesive properties of soot, soot particles can agglomerate into cluster sizes of > 500 nm, although the average size of agglomerated soot particles is closer to ~200 nm [102–105].

1.3.1.2 Morphology, structure, and composition of soot

According to statistics on the pollutant emissions of motor vehicles, including high-speed automobiles (e.g., minibuses, medium buses, large buses, minivans, light trucks, medium trucks, and heavy trucks), low-speed automobiles, and motorcycles, the number of soot particles emitted by trucks is significantly greater than that emitted by buses, with heavy trucks being among the main contributors. Therefore, it is very important to study the soot produced by petrochemical diesel relative to potential alternative fuels.

Recently, researchers have investigated the physical and chemical properties of soot. For example, Clague et al. [106] reported that although carbon black (CB), a soot simulant, has the same primary particle size as diesel soot, it shows considerable differences in terms of its element composition and surface chemical functional groups. Using various characterization techniques, Sharma et al. [107] compared the structure and chemical composition of diesel soot extracted from crankcase lubricating oil and soot emitted from an engine exhaust pipe. The main elements of the soot emitted from the engine exhaust pipe were C and O, while the diesel soot extracted from the crankcase lubricating oil also contained Ca, Zn, P, and S. The results of Uy et al. [108] demonstrated that the primary particle size distribution of diesel soot is in the 25–40 nm range. The morphology and chemical composition of diesel soot are closely related to the polarity and hardness of the soot particles. Zeidan et al. [109] reported that the agglomeration of soot particles in lubricating oil is one of the main factors leading to the excessive increase in viscosity of lubricating oil. Mohankumar et al. [110] reported that certain liquid substances and hydrocarbons adsorbed onto the surface of soot emitted by combustion engines.

Diesel soot is primarily composed of C, as well as small amounts of O, S, N, H, and metallic elements. Esangbedo et al. reported that the proportion of C in soot is 80%–90%, while the content of other elements is relatively low. However, these few elements are primarily concentrated on the surface of the soot, which contributes to the polarity of the soot particles [111]. Soot is primary composed of spherical nanosized primary particles (20–50 nm), which aggregate into irregularly shaped secondary particles. The internal nanostructure of soot is mostly a graphite vortex layer structure or an onion ring structure, while some particles have obvious kernel-shell structure [112]. Patel et al. systematically studied the morphology, structure, and chemical composition of soot in diesel engine sump oil, and reported that there was C, Ca, S, P, and O in the soot, in addition to amorphous zinc compounds and calcium sulfate [113]. When compared with CB, soot had many similarities and some differences, both of which demonstrated that they comprise both ordered and disordered carbon. Furthermore, Patel et al. used X-ray absorption near-edge structure spectra (XANES) and high-resolution field emission electron transmission microscopy (HRTEM) images to analyze the chemical composition of soot extracted from in-service diesel engine oil. They demonstrated that the oil contained phosphates and sulfates of Zn and Ca.

In terms of three-body wear, hard nanoparticles, such as $Ca_3(PO_4)_2$ and Fe_2O_3, were observed to be wrapped in the vortex layer structure of soot to form abrasive particles [114]. Wilson et al. studied the soot in rapeseed oil methyl ester, soybean oil methyl ester, mixed oil (50% soybean oil methyl ester and 50% animal oil), and diesel combustion flue gas. They reported that the particle size of biodiesel soot (BDS) is significantly less than that of diesel soot produced under the same experimental conditions. They reported that the BDS had a high proportion of graphite containing a core-shell structure, while the diesel soot contained short, disconnected, nonconcentrically oriented graphene fragments [115].

The working condition of a diesel engine affects the characteristics of its soot. For example, Lu et al. studied the soot produced by an exhaust diesel engine under five working conditions, including its primary particle size and nanostructure, and reported that for different speeds and loads, the primary particle diameter was, on average, 23.8–28.5 nm. They demonstrated that the nanostructure of soot can reflect its particle-formation process, where high loads and low speeds produce large, irregularly sized particles that show evidence of precursor particles, indicating the combined effect of carbon particle aggregation and surface growth [112]. Generally, primary particles with a core-shell structure are formed at different speeds under high loads; however, if the gas in the cylinder has a low temperature and the combustion duration is short, the particles that form at low loads and high speeds show disordered structures. The air–fuel ratio affects both particle size and nanostructure; however, the rotation speed mainly affects particle size. The degree of graphitization of primary particles is affected by the load and has nothing to do with the rotation speed [112].

Because of their high surface free energy, soot particles have a strong thermodynamic tendency to spontaneously coalesce in oil. Soot in oil primary manifests as chain-like or network aggregates. For example, La Rocca et al. studied the characteristics and agglomeration properties of soot in the sump oil of a high-pressure common rail direct-fired diesel engine, and reported that about half of the soot (54%) in the oil existed in a long-chain agglomerate with an average length of 130 nm and a width of 50 nm. Clustered agglomerates with medium branches accounted for the remaining 46%, whose average aspect ratio was 2.5 [116]. The distribution of the diameters of the primary particles was Gaussian in the range of 10–35 nm, with an average particle size of 20.2 nm. All the primary particles exhibited a kernel-shell structure, with several cores of about 4 nm forming a core with a diameter of 8–15 nm and a shell thickness of 4–12 nm. La Rocca et al. used a nanoparticle tracking analysis (NTA) to characterize the agglomeration structure of the soot particles in the sump oil of a lightly loaded diesel engine, the results of which were compared with TEM results. Both technologies demonstrated that the average size of the soot aggregates in the oil was 120 nm [117].

In fact, because of the different compositions and performances of biodiesel and petrochemical diesel, the physical and chemical properties of the soot they produce are obviously different. Hence, it is necessary to conduct an in-depth study on the soot produced by the combustion of biodiesel relative to that produced by petrochemical diesel. Currently, studies on BDS are attracting attention from researchers worldwide. Savic et al. [118] investigated the soot generated by combusting different proportions of biodiesel and petrochemical diesel. Their results demonstrated that with a change of biodiesel content, the particle size of the primary soot particles, and also the length, curvature, and spacing of the graphite layer contained in the soot shell, were significantly different. Moreover, there were a large number of amorphous carbon atoms in the soot cores, which had a low degree of crystallization. Wang et al. [119] studied the soot produced by the combustion of biodiesel prepared from jatropha curcas oil, cottonseed oil, and rubber seed oil. Their results demonstrated that the difference in the saturation and oxygen content of the biodiesel led to a change in the BDS produced by combustion. Mei et al. [105] compared the surface morphology and composition of 0 × diesel soot and BDS using scanning electron microscopy (SEM). Note that the $0^{\#}$ diesel soot particles were larger than the BDS particles, and the arrangement of the former was looser than the latter. Only S was detected in the $0^{\#}$ diesel soot particles.

1.3.2 Soot formation mechanism

1.3.2.1 Fossil fuel soot

Diesel engines are more likely to produce soot compared to gasoline engines because of differences in their fuel injection and ignition methods. In a gasoline engine, fuel is injected during the intake stroke and then a spark is ignited. However, the fuel in a diesel engine is injected during the compression stroke and then compressed. Gasoline engines burn more fully and completely than diesel engines. At present, most diesel engines use direct injection and combustor eddies to help fuel the mixture. From the injection point of view, combustion occurs very rapidly in the form of flame diffusion. Both air and fuel mix very thoroughly; however, during the mixing, the amount of fuel required is excessive, resulting in a high concentration of soot. After diffuse combustion, the combustion continues to slowly consume most of the remaining fuel in the combustion chamber in the form of pyrolysis. This slow-burning process produces additional soot particles and unburned hydrocarbons [120–122]. Throughout the combustion process, soot particles are produced and then destroyed via soot oxidation. Soot oxidation refers to the oxidation reaction of soot or primary particles in contact with various oxides, which reduces the number of soot particles produced. When the reaction occurs, the hydrocarbons wrapped in the soot burn away and the particle size decreases. Through the diffuse-combustion phase of the combustion process, carbon particles are produced during the early stage of the soot-production process with a higher air–fuel ratio and more carbon particles are oxidized. The deep oxidation of soot is one of the most effective means to reduce soot emissions; therefore, catalytic oxidation and the more complete combustion of soot can reduce particulate emissions [123, 124].

To reveal the formation mechanism of soot, researchers have studied its formation process. Omidvarborna et al. [125] summarized previous studies to date and reported that the production process of soot is generally divided into three stages: the growth of polycyclic aromatic hydrocarbons (PAHs), the surface growth of soot particles, and the agglomeration of particles. Torkashvand et al. [126] reported that the fuel first cracks to form

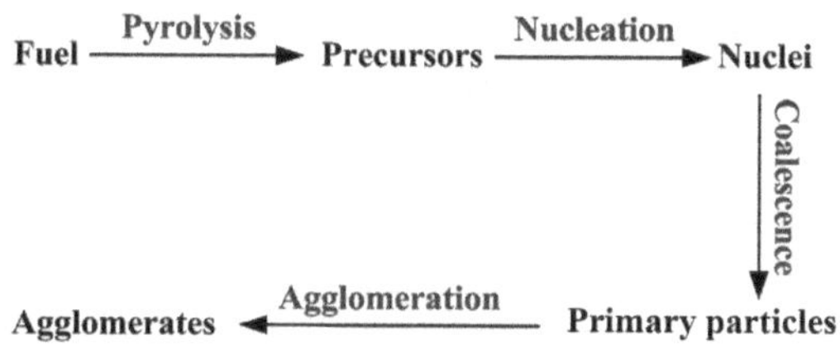

Fig. 1.11 Schematic diagram of the steps in the soot formation.

a precursor. Based on the continuous growth of the precursor, the soot is nucleated and then transformed into a particle through surface growth; the particles further agglomerate to form soot aggregates. This generation process is shown in Fig. 1.11. Kontou et al. [127] suggested that the primary particles of soot agglomerate to form secondary particles, while the secondary particles further agglomerate to form a tertiary structure. This study was helpful for explaining the formation mechanisms of soot. However, the type of fuel and combustion conditions have an impact on the formation of soot; therefore, the relevant mechanisms need to be explored further.

The incomplete combustion of diesel oil in a cylinder produces soot, the amount and properties of which are related to the temperature, oxygen concentration, fuel type, and air–fuel ratio of the diesel oil. When the temperature is high, the oxygen content is low, and the mixture of fuel and air is uneven, fuel cracking occurs easily, producing a lot of soot. Along these lines, Dale et al. studied the process of soot generation in compression–combustion diesel engines and reported that soot was primarily produced in the oil-rich combustion area [128]. Through experiments and simulations, Cheng et al. proposed an improved model for the formation of soot by diesel engines. They demonstrated that during the initial stage of combustion, fuel polymerization and cracking produced a large number of tiny soot particles (5–40 nm). During the intermediate stage of combustion, soot particles continued to grow in size through particle condensation, surface growth, and PAH deposition, which greatly expanded the particle size range and produced some large particles in the engine cylinder. At the end of combustion, the particle size distribution range was stable at 5–20 nm under the influence of further oxidation [129]. After its formation, soot continues to burn, so that part of it is burned off. The temperature of the central area of the cylinder is sufficiently hot enough so that soot combustion can occur. However, in areas near the cylinder wall the temperature is relatively lower, so soot combustion cannot occur. Thus, diesel engine soot is mainly produced in the area near the cylinder wall.

1.3.2.2 Biofuel soot

As previously mentioned, the main component of biodiesel is fatty acid methyl esters, while the main components of diesel are alkanes and aromatic hydrocarbons; these different components result in differences in the composition, morphology, and structure of the soot produced via the combustion of these two fuels. The flame temperature of biodiesel is higher than that of diesel, which is favorable for chemical reactions and which causes the particle sizes of the carbon smoke to be smaller. Biodiesel has a high content of oxygen that can promote combustion, relieve hypoxia, and reduce the formation of soot. Moreover, biodiesel contains almost no aromatic hydrocarbons, and the sulfur content and cetane number are higher than those of diesel fuel; collectively, these properties result in biodiesels generating less soot than diesel [130–132].

In addition to biodiesel, biomass fuel is an important new type of biofuel. Hu et al. improved and emulsified biomass cracking crude oil to prepare biomass fuel [133]. The oxygen-rich content of biomass crude oil is its most important characteristic, which results in many differences between its physical and chemical properties compared to those of fossil fuels. The formation process of soot particles mainly includes the following stages: fuel cracking, formation of precursors (such as aromatic hydrocarbons and PAHs), growth of PAHs, soot nucleation, maturation of the nuclear particles, collisions between particles, the surface growth of soot particles, and the aggregation of soot particles; note that oxidation accompanies the whole process.

Biomass fuel contains ethers, esters, aldehydes, ketones, phenols, alcohols, and organic acids, among which the content of oxygen and aromatics is significantly higher than in diesel. During the fuel-cracking stage, diesel is primarily cracked to form small-molecule active groups, which then form monocyclic aromatics; the proportion of monocyclic aromatics in biomass fuel is $>15\%$. The addition of biomass fuel promotes the generation of additional soot particles, and the possibility of collisions between particles increases, which will eventually lead to the formation of more and larger agglomerated particles. The specific surface area of the soot produced by biomass burning is slightly larger than that of diesel soot.

The flame temperature increases first and then decreases with an increase of biomass fuel content. When the content of refined biomass fuel is 10% and 20%, the calorific value of the fuel decreases slightly, by 1.1% and 6.3%, respectively. An increase of oxygen content in the fuel, on the one hand, promotes combustion; on the other hand, it alleviates the influence of insufficient oxygen in the internal flame. When the content of biomass fuel oil is 20%, the flame temperature is the highest, which promotes chemical reactions; moreover, the diameters of the soot's primary particles are the smallest. Furthermore, the higher flame temperature causes some of the C–OH content to convert into C=O or C–O–C functionals [134]; the unburned fuel contains these components, which are largely adsorbed on the surface of the soot, resulting in a larger content of these functional groups in biomass fuel soot. However, when the refined biomass fuel content is 30%, the calorific value of the fuel significantly decreases (16.5%); at this time, the flame temperature decreases, and the amount of unburned components that adsorb on the surface of carbon smoke increases, among which complex oxygen-containing components and PAHs may increase the degree of graphitization [135].

In terms of their micromorphologies, the soot of biomass fuel and diesel fuel forms chain-like agglomerations. The primary particles have an almost spherical structure, and are connected with each other to form a chain structure. The diameter of the primary particles in 20% emulsified oil soot is the least, i.e., 32 nm. The other primary particles in emulsified oil soot (0%, 10%, and 30%) have diameters of ~38 nm. There is a certain amount of microcrystalline graphite in the soot of both biomass fuel and diesel fuel. The graphitization degree of biomass fuel soot is higher than that of diesel fuel, whereas that of soot is the highest when the content of refined biomass fuel is 30%. Biomass fuel soot and diesel fuel soot are composed of the same elements, and their surfaces contain almost the same functional groups (such as C–OH and C=O). Biomass-fuel soot contains C–O–C functional groups, and when the content of the refined biomass fuel is 20%, the content of C–O–C and C=O groups on the surface is higher than in other soot, which may be attributed to the moderate content of refined biomass fuel in biomass fuel and the high oxygen content of fuel oil. At this time, the flame temperature is the highest, and some C–OH groups are oxidized to both C–O–C and C=O groups [136].

1.4 Hazardous effects of soot

1.4.1 Effect of soot on air pollution

With the expansion of urban areas and the rapid increase of urban populations, the number of motor vehicles in China has rapidly increased, reaching 310 million in 2017, among which 217 million were automobiles (although only 1.53 million were new energy vehicles). This increase in motor vehicles has not only increased economic growth but also has led to increased emission of pollutants. According to the Ministry of Ecology and Environment, the total emission of the four main pollutants from motor vehicles in 2017 was estimated to be 43.597 million tons. This includes 33.273 million tons of carbon monoxide, 4.071 million tons of hydrocarbons, 5.743 million tons of nitrogen oxides, and 509,000 tons of particulate matter. Automobiles have been the main contributor to vehicle air pollution emissions, emitting > 80% of all CO and hydrocarbon pollution, and > 90% of nitrogen oxide and particulate matter pollution. Note that most motor vehicles are driven in densely populated areas, and these exhaust emissions directly harm people's health.

Soot particles are produced during the process of combustion, and these particles are the primary constituent of air pollution, which not only affects the global climate but also harms human health. Epidemiological studies have demonstrated that air pollution is closely related to the occurrence of asthma, cardiovascular diseases, lung cancer, chronic bronchitis, and other diseases. PM2.5, which has diameters equal to or less than 2.5 μm, are particularly harmful to humans, because they are very toxic and easily inhaled [137–141].

1.4.2 Effect of soot on lubricating oil

Soot in lubricating oil accelerates an increase of oil viscosity and oxidation of the lubricating oil, decreases the oil's alkali value, increases its acid value, increases engine wear and engine corrosion, and leads to the formation of sediments and sludge [142, 143]. Among these, heavy load diesel engine soot especially reduces the service life of diesel engine oil, produces a lot of waste oil, and contributes considerable amounts of pollution to the environment. Moreover, soot causes serious wear on parts that experience a lot of friction in heavy-duty diesel engines, which shortens their service lives and leads to the scrapping of diesel engines.

The thickening effect of soot on lubricating oil is a problem for heavy-duty diesel engines. To reduce the emission of particulate matter and NO_X by diesel engines, exhaust-gas recirculation (EGR) technology and delayed injection technology are used in heavy-duty diesel engines, which leads to an increase of soot content in engine oil. Soot pollution in lubricating oil changes the chemical properties of the oil, resulting in a failure of lubricating function, an increase in lubricating oil viscosity, and increased difficulty in pumping the lubricating oil. For example, Penchaliah et al. studied the effects of soot, oxidation, water, and sulfuric acid pollution on the performance of lubricating oil. Both oxidation and soot pollution led to a significant increase in the viscosity of lubricating oil. The wear rate was primarily affected by acid and soot, and the friction coefficient rose when all pollutants were added. Their wear mechanism analysis demonstrated that soot and oxidation caused abrasive wear and polishing wear, respectively, while sulfuric acid and water caused corrosion wear [144]. Next, Brdasz et al. studied the thickening effect of lubricating oil on a Mack engine and reported that the viscosity of lubricating oil increased with an increase of soot content in certain cases and decreased with an increase of soot content in other cases. They found that the viscosity of their oil samples was different depending on the size distribution of the soot; they subsequently hypothesized that the viscosity was directly related to the surface area, content, and particle size distribution of the soot in the oil [145]. The aggregation of primary particles at 30–50 nm in oil can significantly increase the size of the agglomerated particles to >500 nm [146]. The latest research further conjectures that the thickening and excessive viscosity growth of lubricating oil are related to the agglomeration of soot particles in oil [147, 148]. The extreme thickening of diesel engine oil is attributed to the agglomeration of ultrafine carbon particles into larger particle agglomerates with particle sizes >1 μm, where the precipitation of large particle agglomerates will increase the generation of oil sludge in engine oil. The reason that ultrafine carbon particles agglomerate into large particle agglomerates is that there are polar groups on the surface of soot because of the oxidation of oil during the normal operation of the engine. In nonpolar oil, ultrafine particles agglomerate into large particles, which means some of their surfaces are exposed to the oil.

Currently, the industry has recognized the complex dependence between the thickening of lubricating oil and soot. Asango et al. studied the effect of the particle size and concentration of carbon nanoparticles on the viscosity of SAE 5W-30 lubricating oil. Their oil samples were taken from used oil from a light truck engine. The average soot particle size was 180–320 nm, and the soot mass concentration was 0%–2%. Their results demonstrated that the dynamic viscosity of oil increased with an increase of soot concentration and a decrease of temperature, where the viscosity of oil at 40°C increased exponentially with an increase of soot concentration, and the viscosity of oil at 90°C increased linearly with an increase of soot concentration, which had a more significant effect on the particle size of oil with high soot concentration [149]. At lower volume fractions (<0.005), the particle size of 1000-nm graphite particles suspended in oil had a similar effect on increasing viscosity as measured for samples of diesel oil containing soot. When the volume fraction was >0.005, the relative viscosity of oil with added graphite was much higher than that of oil with added diesel soot. The high viscosity of graphite oil samples with a high volume integration rate may be attributed to agglomeration, where the effect of agglomeration may lead to thickening of the lubricating oil and an increase of its relative viscosity. The Stokes–Einstein viscosity equation is suitable for describing oil samples whose soot integral rate is <0.013, while Brinkman's equation is suitable for all oil samples under all experimental conditions.

The different physical and chemical properties of soot will inevitably lead to the obvious difference of its dispersion performance in lubricating oil. For example, Esangbedo et al. studied the morphology, nanostructure, and oxygen functional groups of the soot particles in oil after performance tests of two types of heavy-duty diesel engines with different EGR levels to explore the factors that caused a loss of viscosity control in lubricating oil. They reported that soot with a high graphitization degree and low oxygen functional group content would cause the oil to lose viscosity control. Moreover, even when the soot content of the oil was low, the viscosity control ability was diminished. The reason is that in heavy-duty diesel engines with EGR technology, unreactive soot is produced, which is not easily adsorbed by dispersant molecules, resulting in agglomeration. Unreactive soot can be divided into two types: (1) graphitized soot (whose structure hinders the action of the dispersant), whose graphitization degree determines the activity of the soot; (2) soot with low surface oxygen content (which hinders the action of the dispersant). Furthermore, it is generally believed that oxygen-containing compounds on the surface of soot interact with dispersants.

It has been reported that increase in oil viscosity due to soot is an important problem in some diesel engines. Attempts to control the viscosity increase due to soot primarily depend on using additives with dispersing functions. For example, Covitch et al. compared and studied various lubricating oil formulations on a Mack T-7 bench and reported that a succinimide dispersant, a dispersive viscosity index improver, and a superbasic sulfonate detergent could effectively control viscosity growth [150]. By adsorbing on the acid sites on the surface of the soot, the basic additives blocked the growth of soot particles, interrupted the agglomeration of soot, and interfered with the thickening of oil products. Andrew et al. studied the effect of a batch of dispersive viscosity index improvers on controlling viscosity growth related to soot. A yield stress test was used to evaluate the flow curve and a good dispersive viscosity index improver could improve the rotational viscosity at the yield point. The vibration method was used to evaluate the elastic modulus and viscosity modulus to evaluate the overall flow behavior. Different viscosity index improvers in the experimental concentration range had significant differences in terms of the dispersion ability and viscosity of soot [151].

To reduce the harm caused by large particles, dispersants are added to lubricating oil. Obviously, dispersants in the oil will be affected by other additives (such as antiwear agents and detergents). George used a statistical method to design experiments to study the influence of soot pollution on the viscosity of engine oil, in which a base oil, dispersant, and zinc dithiophosphate (ZDP) were taken as two levels, respectively, and the soot was taken as three levels, respectively, at 40°C and 90°C to study the influence of various factors on oil viscosity [147]. The results demonstrated that the viscosity of the oil increased with an increase of soot content at 40°C and 90°C, while the viscosity changed nonlinearly with an increase of soot concentration at 40°C and linearly at 90°C. Using the general linear model analysis results in a statistical analysis package, the influence of base oil and ZDP could be ignored at 40°C, while the dispersant level and soot water could influence the viscosity of the oil sample at the two temperatures [147].

At present, Mack T-11 engine bench tests are used in industry to control the viscosity increase of oil products caused by soot. The Mack T-11 test stipulates that the viscosity growth at 100°C shall not exceed 12 cSt when the soot content is 6 wt%. For example, David et al. compared a Mack T-11 experimental engine test with a vehicle driving test to try to confirm the relationship between them. The Mack T-11 test provided an effective screening tool to study viscosity growth because of soot. They demonstrated that bench tests are stricter than driving tests, and the Mack T-11 test limit gives a large safety factor [152, 153].

The addition of dispersants in a lubricating oil system can effectively inhibit the agglomeration of soot particles. Through a study of a lubricating oil dispersant system, it was considered that the adsorption of dispersants and soot is the key to determining the dispersion performance of the system. The size of soot particle aggregates is directly related to the dispersant's type and structure. The relationship between dispersant and soot was studied using the molecular simulation method of BLEDN. The dispersing performance of the dispersant increased with the polarity of the polar end of the dispersant. When the interaction between the nonpolar end of the dispersant and lubricating oil was considerable, the dispersion performance was good [154]. The binding energy of polar groups may determine the adsorption strength on the surface of the soot and the performance of the dispersant [155]. Esangbedo et al. reported that polar atoms or groups containing oxygen on the surface of the soot participate in the interaction between both soot and dispersant. Therefore, the oxygen-containing groups on the surface of soot are the adsorption points for the dispersant; moreover, the mechanism of action between the dispersant and soot is based on acid–alkali interactions. The acid functional group on the surface of the soot contains oxygen atoms; thus, the oxygen content can reflect the ability of soot to adsorb dispersants.

In the field of lubricating oil, emphasis is given to the stable dispersion of soot particles. The mechanism of polyisobutylenes (PIBs) was studied in some model solvents. It was found that van der Waals attraction was the cause of agglomeration of the colloidal soot particles, while the stable dispersion was due to electrostatic repulsion or steric hindrance. To understand the dispersion mechanism of soot in oil, Ruot et al. studied the influence of dispersants on the dispersion stability of CB particles in base oil in adsorption and sedimentation experiments. An adsorption isotherm was used to explain the mechanism of dispersion as a function of agglomeration. The dispersant could stabilize the particles in the oil after the critical concentration was exceeded [156]. Moreover, the adsorption of nondispersive polymethacrylate (PAMA) and nondispersive polyolefin (OCP) on CB was limited, and it was not sufficient to protect particles from agglomeration. Dispersed polymers such as DPAMA, DOCP, and polyisobutylene succinimide (HPSB) can be strongly adsorbed on CB to produce repulsive protective particles against agglomeration, which allows the CB particles to disperse and become suspended in oil. Therefore, these additives can minimize the influence of soot in lubricating oil [156]. The coverage of a particle's surface is directly related to the affinity of additives to the surface. This affinity determines whether the additive can be adsorbed on the particle's surface. The effect of a dispersant in oil is related to the proportion of the dispersant completely covering the particle's surface. The actual surface area that CB can provide for additive adsorption depends on the relative molecular size of the additive and the basic CB particle size. Therefore, whether or not the additive can be adsorbed onto CB depends on the molecular weight of the additive rather than the number of polar functional groups. Moreover, it depends on the structure of the adsorption layer. A unique adsorption layer is sufficient to effectively repel the agglomeration of particles.

The molecular structure of a surfactant has an important effect on the dispersion of soot particles in hydrocarbon solvents. For example, Tomlinson et al. studied the effect of surfactant molecular structure on the dispersion of carbon particles in hydrocarbon solvents through an adsorption experiment of succinimide and activated carbon [157]. Both the length and framework of the hydrophobic tail of the dispersant are the primary factors determining the adsorption effect, which may be where the dispersion mechanism lies. Moreover, all succinimides have high affinity for adsorbents. Next, Shen et al. studied the stable dispersion effect of a series of nonionic dispersants on carbon-rich particles in oil and reported that the cross-linking strength of dispersants increased with an increase in the number of secondary amines in the polyamine core, because there are many amines on the polar nucleus of the dispersant [158]. The measured CB adsorption isotherms demonstrated that there were two binding mechanisms: when the dispersant concentration was lower than the CMC concentration, strong bonding occurred, and when the opposite situation occurred, weak bonding occurred, and the dispersant was competing between the micelles and adsorption. Yasin et al. studied the dispersion effect of three dispersants (Hypermer LP1, Hypermer B246, and OLOA 11000) on graphite-type CB in an organic solvent by measuring the rheological property, conductivity, and adsorption capability. The measured adsorption isotherms showed that the adsorption affinities of OLOA 11000 and Hypermer B246 were considerably stronger than that of Hypermer LP1. This may be because of the block polymerization structure of the two former dispersants, while Hypermer LP1 is a homopolymer [159]. In nonaqueous media, Hypermer B246 and OLOA 11000 produce a low-viscosity dispersion system, which is an effective stabilizer, while Hypermer LP1 produces a high-viscosity suspension system, which is not an effective stabilizer. Hypermer LP1 cannot stabilize the suspended graphite CB because of the basic structure of the dispersant; the structure of the surfactant controls its structure on the surface of soot, and Hypermer LP1 has a planar structure because of its adsorption affinity on all surfaces, resulting in an insufficient thickness of the stable layer [159].

The properties of base oil have a significant effect on the ability of the dispersant to adsorb soot particles. For example, Bezot et al. studied the agglomeration kinetics of colloidal CB ions in different types of base oils (group I and group II) using a light-scattering technique. CB had different stabilities in the group I and II base oils: it did not agglomerate in group I base oils, but it obviously agglomerated in group II base oils. Dispersants can slightly reduce the size of agglomerates in group I base oil, while in group II base oil, agglomeration slows down or stops, depending on the type of dispersant [160].

1.4.3 Effect of soot on engine parts

Since the 1980s, research on engine soot wear has been extensive and in-depth, demonstrating that soot particles are related to diesel engine wear. For example, Rounds et al. considered CB in an engine soot simulator and demonstrated that CB had the ability to absorb zinc dialkyldithiophosphates (ZDDPs), which led to the formation of the antiwear layer being blocked, thus aggravating engine wear [161]. Subsequently, Yamaguchi et al. investigated the relationship between the decomposition of ZDDP because of soot in two diesel engine oils and the wear of distribution valves [162]. Their results demonstrated that one oil produced high valve wear and the other produced low valve wear. P-31 nuclear magnetic resonance (NMR) spectra of the oil samples demonstrated that lower valve wear produced more ZDDP decomposition. Rounds believes that engine oil failure caused by soot is due to soot-adsorbing ZDDP, which reduces its content and thus causes valve wear [161, 163]. This explains the necessity of studying soot inside the engine. Nagai et al. proposed that soot has a grinding effect on the antiwear layer formed by oil, thus destroying the antiwear layer. Moreover, they reported that by adjusting the engine structure or the oil formula, it is possible to effectively reduce distribution valve wear [164]. This led to a debate among later scholars about the wear mechanism of soot. Currently, certain researchers are still inclined to the theory of chemisorption in studies of the soot mechanism and have put forward various wear models [165, 166]. For example, Steffen et al. studied the tribochemical characteristics of soot using a pin-disk friction test machine and a high-resolution wear test system (RNT) [167]. High-resolution transmission electron microscopy (HRTEM), electron energy loss spectroscopy (EELS), electron spin resonance (ESR), and photoelectron spectroscopy (XPS) were used to observe the morphology of soot particles. Atomic force microscopy (AFM) and auger electron microscopy (AES) were used to analyze the surface topography of the friction pair.

Because of the complex working conditions of an engine, the lubrication conditions of the valve group, which is most seriously affected by soot wear, are varied. This directly affects the wear mechanism of soot. The research report from SAE indicated that the wear rate is relatively small when the oil film thickness is larger than the original carbon particle diameter. However, when the diameters of the carbon particles are close to the thickness of the oil film, the amount of wear intensifies. Spikes reported that, under the condition of point contact elastohydrodynamic lubrication, soot could reduce the thickness of the lubricating ZnDTP film on the friction surface and the carbon particles themselves played a role in friction reduction [168]. Colacicco analyzed the influence of soot particle deposition on valve wear and reported that soot particles aggregate at the oil inlet of the cam–ejector friction pair, resulting in dry friction at the contact and adhesive wear [169]. Gautam et al. conducted experiments on the wear mechanism of soot. A three-body wear machine was used to simulate and estimate the wear of soot particles [170].

Kontou et al. studied the effect of soot on the friction and wear behavior of ball disk friction pairs. Both soot and ZDDP in the lubricating oil could not reduce the soot wear by increasing the hardness of the disks [171]. Olomelehin et al. found that CB (the simulant of soot) could antagonize the antiwear agent in lubricating oil and destroy the friction reaction film that forms on the surface of a friction pair, resulting in a decrease of the lubricating performance of the lubricating oil [172]. Salehi et al. reported that with an increase of soot content, the wear scar diameter and wear volume significantly increased [173]. Under the same soot content, the wear degree at 100°C was higher than at 40°C. Moreover, they put forward a corrosion-wear mechanism. They suggested that, in the presence of soot, the friction reaction film generated by the full formula lubricating oil is repeatedly scraped off by the soot. The formation speed of the friction reaction film is an important factor affecting the speed of corrosion wear.

Su et al. investigated the effects of diesel soot on the tribological properties of diesel engine oil using ball disks and cylinder liner piston ring friction pairs [174]. Their results demonstrated that for the ball disk friction pairs, when the soot content was at a low level, the soot can reduce the friction coefficient. For the cylinder liner piston ring friction pairs, the soot content, friction conditions, and the degree of wear were closely related. Next, Liu et al. studied the tribological behavior of biomass fuel soot in lubricating oil using a four ball friction and wear test machine. During the process of friction, biomass fuel soot adsorbed on the interface of the friction pair, which limited the lubricating function of the boundary lubrication film [175]. Furthermore, the soot damaged the friction surface and generated abrasive particles, resulting in severe wear. Hu et al. studied the tribological properties of CB in a lubricating oil base oil and a full-formula oil [176]. Their results demonstrated that the size of the CB aggregate was closely related to its tribological mechanism. When the size of the CB aggregate was less than the thickness of the oil film, it played the role of ball lubrication. However, when the size of the CB aggregate was larger than the thickness of the oil film, which led to the friction interface being scraped by the aggregate, the wear increased. Green et al. analyzed the relationship between the friction speed and the friction and wear mechanism of soot-containing lubricating oil [177]. At low speeds, the lubricating oil functioned properly between the friction pairs; at medium speeds, the surface of the friction pairs was scratched by the soot; and at high speeds, the accumulation of soot in the contact area of the friction pairs led to poor oil wear.

To control the effect of soot on the tribological behavior of engine parts, it is necessary to identify an effective approach. Because of the accumulation of charge on the surface of soot particles, they tend to agglomerate in lubricating oil, which results in an increased particle size of the soot agglomerates and an increase in the viscosity of the oil products. These factors lead to the circulation of lubricating oil and the formation of a poorly lubricating oil film, resulting in the increased wear of engine parts and even catastrophic failure. To prevent the agglomeration of soot particles, chemical additives (such as dispersants and detergents) are usually added to lubricating oil to meet the performance requirements of the engine. Along these lines, Covitch et al. compared and analyzed different lubricant additive formulations using a Mack T-7 engine bench [150]. Their test results demonstrated that dispersants and high-alkali detergents could effectively control the viscosity growth of the lubricant, which was primarily attributed to the adsorption of the previously mentioned additives on the surface of soot particles, thus blocking the agglomeration of soot and delaying the thickening of the lubricant. Esangbedo et al. reported that the graphitization degree and surface oxygen content of soot directly affected the adsorption effect of dispersants on soot [178]. Ruot et al. reported that HPSB dispersants and polymers with dispersant function could adsorb on the surface of soot or CB particles, thus preventing the particles from coalescing and facilitating their suspension in lubricating oil [156]. Liu et al. added a succinimide dispersant and calcium detergent to lubricating oil containing biomass fuel soot and reported that the viscosity of the lubricating oil significantly decreased, and the particle size of the agglomerated soot also significantly decreased [179, 180]. Therefore, it is necessary to study the effect of additives on the agglomeration of BDS and the dispersion of BDS on the tribological behavior of lubricating oil.

Adding a proper amount of solid lubricating particles to lubricating oil can effectively improve its lubricating performance. Graphene, as a new two-dimensional material studied extensively in recent years, has great potential as a lubricating material. For example, Gupta et al. reported that the functional groups on the surface of reduced graphene oxide (RGO) significantly affect its friction and antiwear performance in lubricating oil [181]. Furthermore, the amount of RGO added to lubricating oil is an important factor related to the tribological performance of lubricating oil. A proper amount of RGO can play a significant role in lubrication, which is attributed to the low shear strength of RGO nanosheets adsorbed on the friction interface. Zhao et al. selected three types of graphene with different peeling degrees as lubricating additives of a polyalphaolefin (PAO) base oil. They revealed the lubricating mechanism of graphene by analyzing its wear debris [182]. Their results showed that the tribological properties of graphene with different degrees of exfoliation in PAO base oil were quite different. This was primarily attributed to the fact that graphene with a high degree of peeling and only a few layers could adsorb on the friction interface; hence, a layered friction film formed under the action of shear force that protected the friction pair. However, graphene with a low degree of peeling and many layers was randomly adsorbed on the friction interface in various forms, which could easily destroy the friction interface during the friction process. Kamel et al. reported that graphene nanoflakes contained in calcium grease could form an adsorption protective film on the friction surface during the friction process, thus effectively improving the friction reduction and antiwear effect of calcium grease [183]. Zheng et al. investigated the tribological properties of lubricating oil containing graphene nanosheets on the textured surface of a titanium alloy. They reported that graphene nanosheets demonstrated graphitization transformation during the friction process [184]. Zhang et al. studied the tribological behavior of modified graphene oxide (GO) in vegetable oil. Their results showed that during the process of friction, the interlayer sliding of modified GO and S involved in the formation of a lubricating film resulted in reduced friction and better antiwear performance of the vegetable oil [185].

To improve the lubricity of graphene, some researchers have suggested that it is necessary to adhere metal or metal oxides onto the surface of graphene. For example, Xu et al. dispersed Co_3O_4 nanoparticles evenly among different GO layers, thus inhibiting the agglomeration of GO [186]. Meng et al. reported that Ag particles with particle sizes of 2–16 nm were introduced to the surface of graphene via supercritical fluid technology, and it was found that nanosilver particles and graphene had a synergistic lubrication effect [187]. Song et al. prepared GO/α-Fe_2O_3 materials using the hydrolysis method. The lubrication effect of the GO/α-Fe_2O_3 in the lubricating oil was better than that of GO or α-Fe_2O_3 added separately, which was attributed to the priority of GO in film formation at the friction interface and the roller effect of α-Fe_2O_3 [188]. Hu et al. added TiF_3 and LaF_3 to lubricating oil containing biofuel soot to investigate the influence of these two kinds of nanoparticles on the tribological characteristics of the soot-containing lubricating oil [189]. Their results demonstrated that the two fluoro-metal compounds could significantly enhance the friction-reducing and antiwear performance of the lubricating oil, and TiF_3 could induce the transformation of soot during the friction process.

The work of the researchers mentioned has advanced our understanding of the tribological characteristics of soot, i.e., solid particles with excellent lubricating performance were added to lubricating oil containing soot to investigate the effect of solid particles on the BDS containing lubricating oil. This work has also helped to reduce the negative impact of soot on engine parts and provide a new way to extend the oil change cycle of engine lubricating oil.

1.5 Summary

The conclusions of this study are summarized in the following paragraphs.

Biodiesel is composed of various long-chain fatty acid alkyl esters, including saturated and unsaturated compounds, and is typically obtained from raw materials such as vegetable oils, animal fats, or waste cooking oils. Biodiesel can be prepared by homogeneous and heterogeneous catalysts. Engines powered by biodiesel have the benefits of reduced particulate matter emission and thus contribute less than petrodiesel engines to the global greenhouse effect.

However, the generation of soot as a byproduct of the combustion process cannot be eliminated completely. The fuel composition is a key factor that decides the morphology, structure, and composition of soot particles. The physical and chemical properties of BDS are significantly different from those of petrochemical diesel soot. Most BDS generated during the combustion process is exhausted; however, some can contaminate the lubricating oil system of the engine and lead to an increase in the viscosity of the lubricating oil and hence increased engine wear. Indeed, the energy savings and service life of engines are significantly affected by the influence of BDS on the lubrication performance of engine lubricating oil.

References

[1] Singh D, Sharma D, Soni SL, Sharma S, Kumari D. Chemical compositions, properties, and standards for different generation biodiesels: a review. Fuel 2019;253:60–71.
[2] Lee J, Lee B, Heo J, Kim H, Lim H. Techno-economic assessment of conventional and direct-transesterification processes for microalgal biomass to biodiesel conversion. Bioresour Technol 2019;294:122173.
[3] Quah RV, Tan YH, Mubarak NM, Khalid M, Abdullah EC, Nolascohipolito C. An overview of biodiesel production using recyclable biomass and non-biomass derived magnetic catalysts. J Environ Chem Eng 2019;7(4):103219.
[4] Li H, Liu F, Ma X, Cui P, Gao Y, Yu M, Guo M. Effects of biodiesel blends on the kinetic and thermodynamic parameters of fossil diesel during thermal degradation. Energ Conver Manage 2019;198:111930.
[5] Ma Y, Liu S, Wang Y, Adhikari S, Dempster TA, Wang Y. Direct biodiesel production from wet microalgae assisted by radio frequency heating. Fuel 2019;256:115994.
[6] Kumar SA, Sakthinathan G, Vignesh R, Banu JR, Almuhtaseb AH. Optimized transesterification reaction for efficient biodiesel production using Indian oil sardine fish as feedstock. Fuel 2019;253:921–9.
[7] Sun S, Guo J, Duan X. Biodiesel preparation from Phoenix tree seed oil using ethanol as acyl acceptor. Ind Crop Prod 2019;137:270–5.
[8] Mardhiah HH, Ong HC, Masjuki HH, Lim S, Lee HV. A review on latest developments and future prospects of heterogeneous catalyst in biodiesel production from non-edible oils. Renew Sustain Energy Rev 2017;67:1225–36.
[9] Shi WY, Li HB, Cheng F, Yang LM. Research progress of preparation methods for novel biodiesel. Chem Eng Oil Gas 2016;45(1):1–7.
[10] Li Y, Zeng D, Fang T. Review of kinetic research on biodiesel production with supercritical methanol. Chem Ind Eng 2013;30(1):67–78.
[11] Saka S, Kusdiana D. Biodiesel fuel from rapeseed oil as prepared in supercritical methanol. Fuel 2001;80(2):225–31.
[12] Madras G, Kolluru C, Kumar R. Synthesis of biodiesel in supercritical fluids. Fuel 2004;83(14):2029–33.
[13] Köse Ö, Tüter M, Aksoy HA. Immobilized candida antarctica lipase-catalyzed alcoholysis of cotton seed oil in a solvent-free medium. Bioresour Technol 2002;83(2):125–9.
[14] Chen JY, Lu XH, Nie Y, Ji JB. Production of biodiesel through transesterification of soybean oil catalyzed by immobilized lipase. J Chin Cereals Oils Assoc 2016;31(12):57–60.
[15] Mao YY, Li WQ, Wang CY, Shi H. Preparation of biodiesel from tung oil catalyzed by immobilized lipase. Chin Oils Fats 2017;42(1):52–5.
[16] Iso M, Chen B, Eguchi M, Kudo T, Shrestha S. Production of biodiesel fuel from triglycerides and alcohol using immobilized lipase. J Mol Catal B: Enzym 2001;16(1):53–8.
[17] Watanabe Y, Shimada Y, Sugihara A, Tominaga Y. Enzymatic conversion of waste edible oil to biodiesel fuel in a fixed-bed bioreactor. J Am Oil Chem Soc 2001;78(7):703–7.
[18] Marchetti JM, Errazu AF. Esterification of free fatty acids using sulfuric acid as catalyst in the presence of triglycerides. Biomass Bioenergy 2008;32(9):892–5.
[19] Al-Widyan MI, Al-Shyoukh AO. Experimental evaluation of the transesterification of waste palm oil into biodiesel. Bioresour Technol 2002;85(3):253–6.
[20] Qin SJ, Sun YZ, Li P, Yao HW, Shi CL. A comparison of biodiesel preparation from soybean oil by using homogeneous catalysts. J Chin Cereals Oils Assoc 2010;25(1):60–3.
[21] Zhang FJ, Jiang ST, Pan LJ, Liu WY. Biodiesel preparation from rapeseed oil with methanol catalyzed by potassium carbonate. J Chin Cereals Oils Assoc 2011;26(3):52–5.
[22] Freedman B, Butterfield RO, Pryde EH. Transesterification kinetics of soybean oil. J Am Oil Chem Soc 1986;63(10):1375–80.
[23] Nye MJ, Williamson TW, Deshpande S, Schrader JH. Conversion of used frying oil to diesel fuel by transesterification: preliminary test. J Am Oil Chem Soc 1983;60(8):1598–601.

[24] Fukuda H, Kondo A, Noda H. Biodiesel fuel production by transesterification of oils. J Biosci Bioeng 2001;92:405–16.
[25] Liu KS. Preparation of fatty acid methyl esters for gaschromatographic analysis of lipids in biological materials. J Am Oil Chem Soc 1994;71(11):1179–87.
[26] Borges ME, Díaz L. Recent developments on heterogeneous catalysts for biodiesel production by oil esterification and transesterification reactions: a review. Renew Sustain Energy Rev 2012;16(5):2839–49.
[27] Chouhan APS, Sarma AK. Modern heterogeneous catalysts for biodiesel production: a comprehensive review. Renew Sustain Energy Rev 2011;15(9):4378–99.
[28] Mo X, Lotero E, Lu C, Liu Y, Goodwin JG. A novel sulphonated carbon composite solid acid catalyst for biodiesel synthesis. Catal Lett 2008;123(1-2):1–6.
[29] Liu ZP, Ou SY, Wang Y, Wen Y. Preparation of biodiesel from waste cooking oil by two-step catalyzed process. Chin Oils Fats 2006;31(5):59–62.
[30] Wang ZH, Wang ZH, Sun XM, Sun GF, Liu GW. Preparation of biodiesel catalyzed by SO_4^{2-}/Fe_2O_3 solid super acid. Chin Oils Fats 2007;32(8):59–62.
[31] Wen Z, Yu X, Tu T, Yan J, Dahlquist E. Synthesis of biodiesel from vegetable oil with methanol catalyzed by Li-doped magnesium oxide catalysts. Appl Energy 2010;87:743–8.
[32] Kansedo J, Lee KT. Transesterification of palm oil and crude sea mango (*Cerbera odollam*) oil: the active role of simplified sulfated zirconia catalyst. Biomass Bioenergy 2012;40:96–104.
[33] Li Y, Zhang XD, Sun L, Zhang J, Xu HP. Fatty acid methyl ester synthesis catalyzed by solid superacid catalyst SO_4^{2-}/ZrO_2-TiO_2/La^{3+}. Appl Energy 2010;87(1):156–9.
[34] Chen H, Wang JF. Biodiesel preparation from transesterification of cotton seed oil by solid acids catalysis. Chin J Process Eng 2006;6(4):572–5.
[35] Wang D, Su YY, Wang H, Wu ZF, Chen F. Effects of reducing high-acid of biodiesel feedstock with high acid value by carbon-based solid acid. Trans Chin Soc Agric Eng 2009;25(11):274–7.
[36] Sani YM, Daud WMAW, Aziz ARA. Activity of solid acid catalysts for biodiesel production: a critical review. Appl Catal, A 2014;470:140–61.
[37] Bancquart S, Vanhove C, Pouilloux Y, Barrault J. Glycerol transesterification with methyl stearate over solid basic catalysts I. Relationship between activity and basicity. Appl Catal 2001;218:1–11.
[38] Martinez SL, Romero R, Lopez JC, Romero A, Mendieta VS, Natividad R. Preparation and characterization of CaO nanoparticles/NaX zeolite catalysts for the transesterification of sunflower oil. Ind Eng Chem Res 2011;50:2665–70.
[39] Xie WL, Huang XM, Li HT. Soybean oil methyl esters preparation using NaX zeolites loaded with KOH as a heterogeneous catalyst. Bioresour Technol 2007;98:936–9.
[40] Zou HS, Li J, Chen WB. Ultrasonic-enhanced preparation of biodiesel via transesterification catalyzed by solid base. J South Chin Univ Techno 2011;39(11):6–11.
[41] Zhang B, Tao ZH. Synthesis of biodiesel catalyzed by NaF/CaO solid base. Ind Catal 2010;18(2):45–8.
[42] Qian WW, Han PF, LV XP. KF/CaO as solid base catalyst for transesterification to biodiesel by ultrasound. J Fuel Chem Technol 2010;38(1):52–6.
[43] Macleod CS, Harvey AM, Lee AF, Wilson K. Evaluation of the activity and stability of alkali-doped metal oxide catalysts for application to an intensified method of biodiesel production. Chem Eng J 2008;135(12):63–70.
[44] Jiang ST, Xu LY, Zhou QL. Preparation of biodiesel from cottonseed oil catalyzed by solid based catalyst. Trans Chin Soc Agri Eng 2011;27(3):254–9.
[45] Xing XP, Dai Y. Study on solid base catalyst for preparation of biodiesel from *Pistacia chinensis* Bunge oil. J Chin Cereals Oils Assoc 2011;26(11):51–4.
[46] Guo DF, Li WM, Pan JB, Shi HY, Cheng H. Synthesis of biodiesel with solid base CaO/MgO as catalyst. Chin J Appl Chem 2007;24(10):1149–52.

[47] Wan T, Yu P, Gong S, Li Q, Luo Y. Application of KF/MgO as a heterogeneous catalyst in the production of biodiesel from rapeseed oil. Korean J Chem Eng 2008;25(5):998–1003.
[48] Zhang L, Jia W. Research on the preparation of biodiesel oil by mixed oxides from hydrotalcite-like compounds. Shanxi Chem Ind 2009;29(6):34–7.
[49] Zhang Y, Wang J, Chen JL. MgO-Al_2O_3 impregnated with potassium acetate used as a catalyst to produce biodiesel from soybean oil. Chin Oils Fats 2007;32(12):48–52.
[50] Li C, Wang XY, Fang ZL, Cheng L. Development on making biodiesel of the solid alkali catalyst with the metal oxide as the carrier. J Chaohu Coll 2014;3:49–52.
[51] Wang H, Wu Y, Jia QY, Tang XD, Zhong T. Preparation of biodiesel by transesterification over Mg-Al hydrotalcite supported catalyst. Chem Eng Oil Gas 2010;39(5):406–10.
[52] Trakarnpruk W, Porntangjitlikit S. Palm oil biodiesel synthesized with potassium loaded calcined hydrotalcite and effect of biodiesel blend on elastomer properties. Renew Energy 2008;33(7):1558–63.
[53] Liu X, Xiong X, Liu C, Liu D, Wu A, Hu Q, Liu C. Preparation of biodiesel by transesterification of rapeseed oil with methanol using solid base catalyst calcined K_2CO_3/γ-Al_2O_3. J Am Oil Chem Soc 2010;87(7):817–23.
[54] Evangelista JP, Gondim AD, Souza LD, Araujo AS. Alumina-supported potassium compounds as heterogeneous catalysts for biodiesel production: a review. Renew Sustain Energy Rev 2016;59:887–94.
[55] Jiang ST, Cai J, Pan LJ, Zhang FJ, Zhang Y. Preparation technology of biodiesel using K_2CO_3 loaded hydrotalcite catalyst. Trans Chin Soc Agric Eng 2009;40(4):102–6.
[56] Sun GD, Liu Y, Zhai LX, Wu MC. Optimization of the technology for biodiesel preparation using solid alkali as heterogeneous catalyst. Trans Chin Soc Agric Eng 2008;24(5):191–5.
[57] Chakraborty R, Bepari S, Banerjee A. Application of calcined waste fish (*Labeo rohita*) scale as low-cost heterogeneous catalyst for biodiesel synthesis. Bioresour Technol 2011;102(3):3610–8.
[58] Viriya-Empikul N, Krasae P, Puttasawat B, Yoosuk B, Chollacoop N, Faungnawakij K. Waste shells of mollusk and egg as biodiesel production catalysts. Bioresour Technol 2010;101(10):3765–7.
[59] Boey PL, Maniam GP, Hamid SA, Ali DMH. Utilization of waste cockle shell (*Anadara granosa*) in biodiesel production from palm olein: optimization using response surface methodology. Fuel 2011;90(7):2353–8.
[60] Corro G, Sánchez N, Pal U, Bañuelos F. Biodiesel production from waste frying oil using waste animal bone and solar heat. Waste Manag 2016;47:105–13.
[61] Chen H, Wang JF. Kinetic of KOH catalyzed trans-esterification of cottonseed oil for biodiesel production. J Chem Ind Eng 2005;56(10):1971–4.
[62] Malhotra R, Ali A. 5-Na/ZnO doped mesoporous silica as reusable solid catalyst for biodiesel production via transesterification of virgin cottonseed oil. Renew Energy 2019;209:606–19.
[63] Diasakou M, Louloudi A, Papayannakos N. Kinetics of the non-catalytic transesterification of soybean oil. Fuel 1998;77(12):1297–302.
[64] Kudiana D, Saka S. Kinetics of transesterification in rapeseed oil to biodiesel fuel as treated in supercritical methanol. Fuel 2001;80:693–8.
[65] Darnoko D, Cheryan M. Kinetics of palm on transesterification in a batch reactor. J Am Oil Chem Soc 2000;77(12):1263–7.
[66] Isigigur A, Karaosmanoglu F, Aksoy HA. Methyl ester from safflower seed oil of Turkish origin as a biofuel for diesel engines. Appl Biochem Biotechnol 1994;45(1):103–12.
[67] Wu GY, Lin YP, Wu MX, Wang L, Zhou YJ. Kinetic of cottonseed oil transesterification in a batch reactor. J Chem Eng Chinese Univ 2003;17:314–418.

[68] Encinar JM, González JF, Rodríguez-Reinares A. Ethanolysis of used frying oil biodiesel preparation and characterization. Fuel Process Technol 2007;88:513–22.
[69] Nabi MN, Rahman MM, Akhter MS. Biodiesel from cotton seed oil and its effect on engine performance and exhaust emissions. Appl Therm Eng 2009;29:2265–70.
[70] Royon D, Daz M, Ellenrieder G, Locatelli S. Enzymatic production of biodiesel from cotton seed oil using t-butanol as a solvent. Bioresour Technol 2007;98(3):648–53.
[71] Huang Y, Li J, Yan YJ. Novozym 435 catalyzed biodiesel production from cottonseed oil with methyl acetate as acyl acceptor. China Oils and Fats 2008;33(2):40–3.
[72] Li H, Shen BX. Technique optimization and mathematical model of biodiesel and glycerol production from cottonseed oil. China Oils and Fats 2008;33(2):50–3.
[73] Li YB. Preparation of biodiesel from cottonseed oil. Henan University of Technology; 2007.
[74] Wang BT, Qian JF, Liu S, Yun Z. Preparation of fatty acid methyl ester from cottonseed oil produced by two-phase solvent extraction. J Chem Ind Eng 2007;58(10):2636–40.
[75] Zhang H, Meng YB. Mass production of bio-diesel from cottonseed oil via transesterification. Chem Ind Eng Prog 2007;26(1):86–8.
[76] Wu GY, Lin XP, Wu MX, Wang L, Zhou YJ. Kinetics of cottonseed oil transesterification in a batch reactor. J Chem Eng Chin Univ 2003;17(3):314–8.
[77] Shu Q, Zhang Q, Xu G, Nawaz Z, Wang D, Wang J. Synthesis of biodiesel from cottonseed oil and methanol using a carbon-based solid acid catalyst. Fuel Process Technol 2009;90(7):1002–8.
[78] Souza AG, Danta HJ, Silva MCD, Santos IM, Fernandes VJ, Sinfronio FSM, Teixeira LSG, Novak C. Thermal and kinetic evaluation of cotton oil biodiesel. J Therm Anal Calorim 2007;90(3):945–9.
[79] Liu YC, Cao WB, Sun CQ. Research on physical and chemical properties of cottonseed biodiesel. Renew Energy 2008;26(6):34–7.
[80] Chen ZZ, Zou LZ, Yuan J, Wu QC, Guo YX. Solubilities of multi-component systems in biodiesel production from cottonseed oil. Chem Ind Eng Prog 2007;26(7):1028–31.
[81] Joshi H, Toler JE, Walker TH. Optimization of cottonseed oil ethanolysis to produce biodiesel high in gossypol content. J Am Oil Chem Soc 2008;85:357–63.
[82] Li T, Zhou JH, Huang M, Liu ZQ. Enzymatic synthesis of biodiesel from rapeseed oil in n-hexane system. Chin Oils Fats 2007;32(12):45–7.
[83] Zhou W, Konar SK, Boocock DGV. Ethyl esters from the single-phase base-catalyzed ethanolysis of vegetable oils. J Am Oil Chem Soc 2003;80:367–71.
[84] Boocock DGB, Konar SK, Mao V, Sidi H. Fast one–phase oil–rich processes for the preparation of vegetable oil methyl esters. Biomass Bioenergy 1996;11(1):43–50.
[85] Choi OK, Song JS, Cha DK, Lee JW. Biodiesel production from wet municipal sludge: evaluation of in situ transesterification using xylene as a cosolvent. Bioresour Technol 2014;166:51–6.
[86] Liu WW, Su YY, Zhang WD, Liu SQ, Xia CF. A study about analysis of glycerin content in biodiesel. Renew Energy 2005;121(3):17–20.
[87] Sharma YC, Singh B, Upadhyay SN. Advancements in development and characterization of biodiesel: a review. Fuel 2008;87:2355–73.
[88] Liu SY, Bao ZH. Effect of cosolvent on transesterification reaction of preparing biodiesel. Chin Oils Fats 2007;32:38–42.
[89] Guan GQ, Nozomi S, Katsuki K. Synthesis of biodiesel from sunflower oil at room temperature in the presence of various cosolvents. Chem Eng J 2009;146:302–6.
[90] Zhang JY, Wu GY, Lin XP, Wei B. Kinetic study of biodiesel preparation with cosolvent. Chin Oils Fats 2008;33:46–8.
[91] Miao X, Wu Q. Biodiesel production from heterotrophic microalgal oil. Bioresour Technol 2006;97:841–6.
[92] Schwad AW, Bagby MO, Freedman B. Preparation and properties of diesel fuels from vegetable oils. Fuel 1987;66:1372–8.

[93] Atadashi IM, Aroua MK, Aziz AR, Sulaiman NM. The effects of catalysts in biodiesel production: a review. J Ind Eng Chem 2013;19(1):14–26.
[94] Chen KT, Wang JX, Dai YM, Wang PH, Liou CY, Nien CW, Wu JS, Chen CC. Rice husk ash as a catalyst precursor for biodiesel production. J Taiwan Inst Chem Eng 2013;44:622–9.
[95] Xie J, Zheng XS, Dong AQ, Xiao ZD, Zhang JH. Biont shell catalyst for biodiesel production. Green Chem 2008;11:355–64.
[96] Jairam S, Kolar P, Sharma-Shivappa R, Osborne JA, Davis JP. KI-impregnated oyster shell as a solid catalyst for soybean oil transesterification. Bioresour Technol 2012;104:329–35.
[97] Suwannasom P, Tansupo P, Ruangviriyachai C. A bone-based catalyst for biodiesel production from waste cooking oil. Energ Source Part A 2016;38:3167–73.
[98] Nisar J, Razaq R, Farooq M, Iqbal M, Khan RA, Sayed M, Shah A, Rahman IU. Enhanced biodiesel production from Jatropha oil using calcined waste animal bones as catalyst. Renew Energy 2017;101:111–9.
[99] Yaakob Z, Sukarman ISB, Narayanan B, Abdullah SRS, Ismail M. Utilization of palm empty fruit bunch for the production of biodiesel from *Jatropha curcas* oil. Bioresour Technol 2012;104:695–700.
[100] Kotwal MS, Niphadkar PS, Deshpande SS, Bokade VV, Joshi PN. Transesterification of sunflower oil catalyzed by flyash-based solid catalysts. Fuel 2009;88:1773–8.
[101] Yu N, Wei YJ, Hu M, Zeng LM, Zhang YH. Characterization and source identification of ambient organic carbon in PM2.5 in urban and suburban sites of Beijing. Acta Sci Circumst 2009;29(2):243–51.
[102] Agarwal AK, Singh AP, Lukose J, Gupta T. Characterization of exhaust particulates from diesel fueled homogenous charge compression ignition combustion engine. J Aerosol Sci 2013;58:71–85.
[103] Chowdhury S, Boyette WR, Roberts WL. Time-averaged probability density functions of soot nanoparticles along the centerline of a piloted turbulent diffusion flame using a scanning mobility particle sizer. J Aerosol Sci 2017;106:56–67.
[104] Trivanovic U, Corbin JC, Baldelli A, Peng WH, Yang JC, Kirchen P, Miller JW, Lobo P, Gagnéb S, Rogak SN. Size and morphology of soot produced by a dual-fuel marine engine. J Aerosol Sci 2019;138:105448.
[105] Mei DQ, Wang Z, Yuan YN, Ruan ZQ. Analysis of the features of the particulate matters generated from diesel engine fueled with biodiesel. Trans Chin Soc Agric Eng 2006;22(12):113–6.
[106] Clague ADH, Donnet JB, Wang TK, Peng JCM. A comparison of diesel engine soot with carbon black. Carbon 1999;37(10):1553–65.
[107] Sharma V, Uy D, Gangopadhyay A, O'Neill A, Paxton WA, Sammut A, Ford MA, Aswath PB. Structure and chemistry of crankcase and exhaust soot extracted from diesel engines. Carbon 2016;103:327–38.
[108] Uy D, Ford MA, Jayne DT, O'Neill A, Haack LP, Hangas J, Jagner M, Sammut A, Gangopadhyay A. Characterization of gasoline soot and comparison to diesel soot: morphology, chemistry, and wear. Tribol Int 2014;80:198–209.
[109] Zeidan M, Jia X, Williams RA, Wedlock DJ. Simulation of aggregation with applications of soot-laden lubricating fluids. Part Part Syst Char 2004;21(6):473–82.
[110] Mohankumar S, Senthilkumar P. Particulate matter formation and its control methodologies for diesel engine: a comprehensive review. Renew Sustain Energy Rev 2017;80:1227–38.
[111] Li C, Zhang QQ, Li ZY, Feng WM, Liu TX, Hu XG. Dispersion of biodiesel soot and its influence on tribological behavior of liquid paraffin. J Chem Ind Eng 2018;69(6):274–82.
[112] Lu T, Cheung CS, Huang Z. Effects of engine operating conditions on the size and nanostructure of diesel particles. J Aerosol Sci 2012;47:27–38.

[113] Patel M, Ricardo CLA, Scardi P, Aswath PB. Morphology, structure and chemistry of extracted diesel soot—part I: transmission electron microscopy, Raman spectroscopy, X-ray photoelectron spectroscopy and synchrotron X-ray diffraction study. Tribol Int 2012;52:29–39.
[114] Patel M, Aswath PB. Morphology, structure and chemistry of extracted diesel soot—part II: X-ray absorption near edge structure (XANES) spectroscopy and high resolution transmission electron microscopy. Tribol Int 2012;52:17–28.
[115] Wilson M, Sergio GS, Stephen M. Analysis of soot particles derived from biodiesels and diesel fuel air-flames. Fuel 2012;102:525–35.
[116] La Rocca A, Di Liberto G, Shayler PJ, Fay MW. The nanostructure of soot-in-oil particles and agglomerates from an automotive diesel engine. Tribol Int 2013;61:80–7.
[117] La Rocca A, Di Liberto G, Shayler PJ, Parmenter CDJ, Fay MW. Application of nanoparticle tracking analysis platform for the measurement of soot-in-oil agglomerates from automotive engines. Tribol Int 2014;70:142–7.
[118] Savic N, Rahman MM, Miljevic B, Saathoff H, Naumann KH, Leisner T, Riches J, Gupta B, Motta N, Ristovski Z. Influence of biodiesel fuel composition on the morphology and microstructure of particles emitted from diesel engines. Carbon 2016;104:179–89.
[119] Wang Z, Li L, Wang J, Reitz RD. Effect of biodiesel saturation on soot formation in diesel engines. Fuel 2016;175:240–8.
[120] Merchan-Merchan W, Sanmiguel SG, McCollam S. Analysis of soot particles derived from biodiesels and diesel fuel air-flames. Fuel 2012;102:525–35.
[121] Jiménez-Espadafor FJ, Torres M, Velez JA, Carvajal E, Becerra JA. Experimental analysis of low temperature combustion mode with diesel and biodiesel fuels: a method for reducing NO_x and soot emissions. Fuel Process Technol 2012;103:57–63.
[122] Nguyen D, Honnery D. Combustion of bio-oil ethanol blends at elevated pressure. Fuel 2008;87(2):232–43.
[123] Matarrese R, Lietti L, Castoldi L, Busca G, Forzatti P. Effect of soot during operation of a Pt-K/Al_2O_3 LNT catalyst. Top Catal 2013;56:477–82.
[124] Tree DR, Svensson KI. Soot processes in compression ignition engines. Prog Energy Combust Sci 2007;33(3):272–309.
[125] Omidvarborna H, Kumar A, Kim DS. Recent studies on soot modeling for diesel combustion. Renew Sustain Energy Rev 2015;48:635–47.
[126] Torkashvand B, Maier L, Hettel M, Schedlbauer T, Grunwaldt J, Deutschmann O. On the challenges and constrains of ultra-low emission limits: Formaldehyde oxidation in catalytic sinusoidal-shaped channels. Chem Eng Sci 2019;195:841–50.
[127] Kontou A, Southby M, Morgan N, Spikes HA. Influence of dispersant and ZDDP on soot wear. Tribol Lett 2018;66:157.
[128] Dale RT, Kenth IS. Soot processes in compression ignition engines. Prog Energy Combust Sci 2007;33:272–309.
[129] Cheng X, Chen L, Yan F, Dong S. Study on soot formation characteristics in the diesel combustion process based on an improved detailed soot model. Energ Conver Manage 2013;75:1–10.
[130] Magín L, Barba J, Sediako AD, Kholghy MR, Thomson MJ. Morphological analysis of soot agglomerates from biodiesel surrogates in a coflow burner. J Aerosol Sci 2017;111:65–74.
[131] Zhang YL, Zhang RL, Rao LZ, Kim DC, Kook SH. The influence of a large methyl ester on in-flame soot particle structures in a small-bore diesel engine. Fuel 2017;194:423–35.
[132] Merchan-Merchan W, Abdihamzehkolaei A, Merchan-Breuer DA. Formation and evolution of carbon particles in coflow diffusion air flames of vaporized biodiesel, diesel and biodiesel-diesel blends. Fuel 2018;226:263–77.

[133] Xiu S, Shahbazi A. Bio-oil production and upgrading research: a review. Renew Sustain Energy Rev 2012;16(7):4406–14.
[134] Hu EZ, Hu XG, Xu YF, Liu YM, Song RH, Yao ZH. Investigation of morphology, structure and composition of biomass-oil soot particles. Appl Surf Sci 2013;271(1):596–603.
[135] Hu EZ, Hu XG, Liu TX, Dearn KD, Xu HM. Role of TiF_3 catalyst in the tribological properties of biofuel soot-contaminated liquid paraffin. Tribol Int 2014;77:122–31.
[136] Li C, Wu B, Liu TX, Liu YM, Hu XG. Physical-chemical and aggregation properties of biodiesel soot particles prepared at normal temperature and atmospheric pressure. Trans Chin Soc Agric Eng 2017;33(13):90–6.
[137] Kim KB, Masiello KA, Hahn DW. Reduction of soot emissions by iron pentacarbonyl in isooctane diffusion flames. Combust Flame 2008;154:164–80.
[138] Miao Z, Baležentis T, Shao S, Chang D. Energy use, industrial soot and vehicle exhaust pollution—China's regional air pollution recognition, performance decomposition and governance. Energy Econ 2019;83:501–14.
[139] Zhang B, Zuo HY, Huang ZH, Tan JQ, Zuo QS. Endpoint forecast of different diesel-biodiesel soot filtration process in diesel particulate filters considering ash deposition. Fuel 2020;272:117678.
[140] Bello YH, Ookawara S, Ahmed M, Elkhouly MA, Elmehasseb IM, Elshafai NM, Elwardany AE. Investigating the engine performance, emissions and soot characteristics of CI engine fueled with diesel fuel loaded with graphene oxide-titanium dioxide nanocomposites. Fuel 2020;269:117436.
[141] Camposeco R, Castillo S, Hinojosareyes M, Nava N, Zanella R. Manganese promoted TiO_2 and ZrO_2 nanostructures for soot combustion with boosted efficiency. Surf Coat Technol 2020;384:125305.
[142] Ratoi M, Castle RC, Bovington CH, Spikes HA. The influence of soot and dispersant on ZDDP film thickness and friction. Lubr Sci 2004;17:25–43.
[143] Jao TC, Li S, Yatsunami K, Chen S, Csontos AA, Howe JM. Soot characterisation and diesel engine wear. Lubr Sci 2004;16:111–26.
[144] Penchaliah R, Harvey TJ, Wood RJK, Nelson K, Powrie HEG. The effects of diesel contaminants on tribological performance on sliding steel on steel contacts. Proc Inst Mech Eng Part J J Eng Tribol 2011;225:779–97.
[145] Bardasz E, Cowling S, Ebeling V, George H. Understanding soot mediated oil thickening through designed experimentation—part 1: Mack EM 6-287, GM 6.2L. SAE Paper 952527, 1995.
[146] Tormos B, Novella R, Gomezsoriano J, Garciabarbera A, Tsuji N, Uehara I, Alonso M. Study of the influence of emission control strategies on the soot content and fuel dilution in engine oil. Tribol Int 2019;136:285–98.
[147] George S, Balla S, Gautam V, Gautam M. Effect of diesel soot on lubricant oil viscosity. Tribol Int 2007;40(5):809–18.
[148] Bardasz E, Carrick V, George H, Graf M. Understanding soot mediated oil thickening through designed experimentation part 4: Mack T-8 test. SAE Paper 971693, 1997.
[149] Asango A, La Rocca A, Shayler P. Investigating the effect of carbon nanoparticles on the viscosity of lubricant oil from light duty automotive diesel engines. SAE Paper 2014-01-1481, 2014.
[150] Covitch MJ, Humphrey BK, Ripple DE. Oil thickening in the Mack T-7 engine test-fuel effects and the influence of lubricant additives on soot aggregation. SAE Paper 852126, 1985.
[151] Smith A G, Lamb G D. Evaluation of dispersant viscosity modifiers for the application of controlling soot-related viscosity increase in EGR-enabled diesel engines. SAE Paper 2007-01-4142, 2007.
[152] David AD, Mark R, Amy LS, Lewis W. Soot related viscosity increase–a comparison of the Mack T-11 engine test to field performance. SAE Paper 2004-01-3009, 2004.

[153] David AD, Mark R, Amy LS, Lewis W. Soot-related viscosity increase–further studies comparing the Mack T-11 engine test to field performance. SAE Paper 2005-01-3714, 2005.
[154] Wei KC, Lu DY, Zhou H, Xu W, Wen H, Xu ZH. Mesoscale simulation for succinimide dispersant systems. Acta Phys Chim Sin 2004;20(6):602–7.
[155] Wei KC, Wen H, Zhou H, Xu W, Xu ZH. The adsorption behaviors of succinimide dispersants on pseudo-soot surface by molecular dynamics simulation. Comput Appl Chem 2005;22(1):11–5.
[156] Ruot CD, Faure D, Blanc G. Adsorption of engine lubricant dispersants and polymers onto carbon black particles. SAE Paper 2000-01-1991, 2000.
[157] Tomlinson A, Scherer B, Karakosta E, Oakey M, Danks TN, Heyes DM, Taylor SE. Adsorption properties of succinimide dispersants on carbonaceous substrates. Carbon 2000;38:13–28.
[158] Shen Y, Duhamel J. Micellization and adsorption of a series of succinimide dispersants. Langmuir 2008;24:10665–73.
[159] Yasin S, Luckham PF, Iqbal T, Zafar M, Ramzan N. Adsorption and rheology of graphitic carbon black nonaqueous dispersions prepared using nonionic surfactants. J Dispers Sci Technol 2013;34(5):737–46.
[160] Bezot P, Hesse-Bezot C, Yamaguchi ES, Harrison JJ, Campbell CB. A light scattering study of the aggregation kinetics of carbon black suspensions in group 1 and group 2 base oils influence of dispersant additives. SAE Paper 2003-01-1998, 2003.
[161] Rounds F. The generation of synthetic diesel engine oil soots for wear studies. Lubr Eng 1984;40(7):394–401.
[162] Yeh SW, Ryason PR, Roby SH, Untermann M, Yamaguchi ES. Soot wear in diesel engines. Proc Inst Mech Eng Part J J Eng Tribol 2006;220(5):463–9.
[163] Rounds FG. Carbon: cause of diesel engine wear? SAE Paper 770829, 1977.
[164] Nagai I, Endo H, Nakamura H, Yano H. Soot and valve train wear in passenger car diesel engines. SAE Paper 831757, 1983.
[165] Godet M. Third-bodies in tribology. Wear 1990;136(1):29–45.
[166] Ryason P, Chan I, Gilmore J. Polishing wear by soot. Wear 1990;137(1):15–24.
[167] Antusch S, Dienwiebel M, Nold E, Albers P, Spicher U, Scherge M. On the tribochemical action of engine soot. Wear 2010;269(1-2):1–12.
[168] Fujita H, Spikes H. The influence of soot on lubricating films. Tribol Ser 2003;43:37–43.
[169] Colacicco P, Mazuyer D. The role of soot aggregation on the lubrication of diesel engines. Tribol Trans 1995;38(4):959–65.
[170] Gautam M, Chitoor K, Durbha M, Summers JC. Effect of diesel soot contaminated oil on engine wear—investigation of novel oil formulations. Tribol Int 1999;32(12):687–99.
[171] Kontou A, Southby M, Spikes HA. Effect of steel hardness on soot wear. Wear 2017;390:236–45.
[172] Olomolehin Y, Kapadia R, Spikes HA. Antagonistic interaction of antiwear additives and carbon black. Tribol Lett 2010;37(1):49.
[173] Salehi FM, Khaemba DN, Morina A, Neville A. Corrosive–abrasive wear induced by soot in boundary lubrication regime. Tribol Lett 2016;63(2):19.
[174] Su P, Xiong Y, Liu X, Yang H, Fan LJ. The influence of diesel soot on tribological behavior of cylinder liner/piston ring. Tribol 2017;37(2):206–11.
[175] Liu TX, Hu EZ, Jin T, Xu YF, Yu HQ, Hu XG. Composition and structure as well as tribological performance of bio-fuel soot. Tribol 2014;34(4):379–86.
[176] Hu EZ, Hu XG, Liu TX, Fang L, Dearn KD, Xu HM. The role of soot particles in the tribological behavior of engine lubricating oils. Wear 2013;304(1-2):152–61.
[177] Green DA, Lewis R, Dwyer-Joyce RS. Wear effects and mechanisms of soot-contaminated automotive lubricants. Proc Inst Mech Eng Part J J Eng Tribol 2006;220(3):159–69.

[178] Esangbedo C, Boehman AL, Perez JM. Characteristics of diesel engine soot that lead to excessive oil thickening. Tribol Int 2012;47:194–203.
[179] Liu TX, Hu EZ, Song RH, Zhang B, Hu XG. Dispersion characteristics of biofuel soot particles. J Chem Ind Eng 2015;66(4):1506–13.
[180] Liu TX, Song RH, Liu YM, Xu YF, Hu XG. Effect of calcium salt detergent on dispersity of biofuel soot. J Chem Ind Eng 2016;67(7):3032–9.
[181] Gupta B, Kumar N, Panda K, Kanan V, Joshi S, Visoly-Fisher I. Role of oxygen functional groups in reduced graphene oxide for lubrication. Sci Rep 2017;7:45030.
[182] Zhao J, Mao JY, Li YR, He YY, Luo JB. Friction-induced nano-structural evolution of graphene as a lubrication additive. Appl Surf Sci 2018;434:21–7.
[183] Kamel BM, Mohamed A, El Sherbiny M, Abed KA, Abd-Rabou M. Tribological properties of graphene nanosheets as an additive in calcium grease. J Dispers Sci Technol 2017;38(10):1495–500.
[184] Zheng D, Cai ZB, Shen MX, Li ZY, Zhu MH. Investigation of the tribology behaviour of the graphene nanosheets as oil additives on textured alloy cast iron surface. Appl Surf Sci 2016;387:66–75.
[185] Zhang GQ, Xu Y, Xiang XZ, Zheng GL, Zeng XQ, Li ZP, Ren TH, Zhang YD. Tribological performances of highly dispersed graphene oxide derivatives in vegetable oil. Tribol Int 2018;126:39–48.
[186] Xu C, Wang X, Zhu JW, Yang XJ, Lu LD. Deposition of Co_3O_4 nanoparticles onto exfoliated graphite oxide sheets. J Mater Chem 2008;18(46):5625–9.
[187] Meng Y, Su FH, Chen YZ. Supercritical fluid synthesis and tribological applications of silver nanoparticle-decorated graphene in engine oil nanofluid. Sci Rep 2016;6:31246.
[188] Song HJ, Jia XH, Li N, Yang XF, Tang H. Synthesis of α-Fe_2O_3 nanorod/graphene oxide composites and their tribological properties. J Mater Chem 2012;22(3):895–902.
[189] Hu EZ, Hu XG, Liu TX, Song RH, Dearn KD, Xu HM. Effect of TiF_3 catalyst on the tribological properties of carbon black-contaminated engine oils. Wear 2013;305(1–2):166–76.

CHAPTER 2

Basic properties of biodiesel soot

2.1 Green biomass catalysts for biodiesel production

2.1.1 Introduction

Biodiesel is made from animal fats and waste cooking oil by transesterification with other short-chain alcohols. It is a mixture of fatty acid methyl esters, making it a useful form of renewable energy. Currently, the industry mainly uses a homogeneous catalyst, such as potassium hydroxide or sodium hydroxide, to synthesize biodiesel. Using a homogeneous catalyst for synthesis has several advantages, including fast response rate and high conversion rate. However, this type of catalyst is difficult to separate from the reacting system, and the posttreatment processes produce large amounts of wastewater. To solve the problems associated with homogeneous catalysts, researchers have developed a variety of heterogeneous catalysts, including solid acid, solid base, and immobilized biological catalysts [1]. However, these catalysts are usually expensive and are not widely used. Thus, it is necessary to carry out additional research on such catalyst materials, to develop a heterogeneous catalyst that has potential industrial application. An important way to reduce the cost of biodiesel production is to use the extensive worldwide biomass resources as catalysts.

2.1.2 Experiment

2.1.2.1 Materials

Commercial soybean oil was obtained from China Oil and Foodstuffs Corporation. Cedar ash was prepared from cedar leaves that were aged and had fallen off naturally. Methanol, phenolphthalein, 2,4-dinitroaniline, 4-nitroaniline, ethanol (95%), sodium thiosulfate, potassium iodide, potassium periodate, potassium dichromate, cyclohexane, sulfuric acid, hydrochloric acid, and soluble starch were all analytically pure.

2.1.2.2 Catalyst preparation

Cedar ash catalyst: First of all, the cedar ash was dried at 110°C for 10 h. The dried cedar ash was first calcined at high temperature. It was then hydrated, dried at 110°C for 12 h, and then placed in a box electric-resistance furnace to be calcined a second time.

Biodiesel Soot
https://doi.org/10.1016/B978-0-12-820553-2.00002-5

2.1.2.3 Characterization of catalyst

Inductively coupled plasma mass spectrometry (ICP-MS, X Series 2, Thermo Fisher Scientific) was used to analyze the main metal element content of cedar ash. Simultaneous thermal analysis (TG, STA449F3, NETZSCH) was also used under a nitrogen atmosphere. A heating rate of 20°C/min for TG analysis was used to study the catalyst and its precursor with change of temperature. X-ray diffractometry (XRD, D/max 2500PC, Japan Science Company) was used to evaluate the catalyst composition and phase change. Field emission scanning electron microscopy (FESEM) (SU8020, Hitachi, Japan) was conducted to analyze the internal structure and size of the catalyst materials and catalyst precursors. The base strength of the catalyst was determined by the Hammett indicator method: the Hammett indicators included phenolphthalein ($H- = 9.8$), 2,4-dinitroaniline ($H- = 15.0$), and 4-nitroaniline ($H- = 18.4$). The Hammett indicators were prepared using a 0.5% methanol solution. The following steps were taken: accurately measure 5 mL cyclohexane, 0.2 g catalyst, and the correct amounts of the preceding indicators in 25 mL tubes; after adsorption equilibrium, the color of the catalyst surface change was observed.

2.1.2.4 Synthesis of biodiesel

The weighted soybean oil was added into a three-necked round bottom flask with reflux condenser and thermometer. According to a certain methanol/oil molar ratio, methanol, soybean oil, and catalyst were placed in a vessel, and the reaction vessel was placed in a water bath under stirring. After the reaction, the catalyst was filtered in vacuum and recovered. The crude biodiesel was separated from the upper layer, and the biodiesel was separated from the crude biodiesel oil by vacuum distillation. Glycerol content was determined by the potassium periodate oxidation method, since the conversion of glycerol reflects the conversion of biodiesel indirectly [2].

The formula for glycerol content (Eq. 2.1) is:

$$w\ (\%) = \frac{(V_0 - V_1) \times M \times C}{4000 \times m} \times 100 \tag{2.1}$$

V_0: the initial volume of $Na_2S_2O_3$ standard solution consumed, mL; V_1: volume of $Na_2S_2O_3$ standard solution consumed in sample, mL; C: concentration of $Na_2S_2O_3$ standard solution, mol/L; M: molar mass of glycerol, g/mol; m: mass of sample, g; w: glycerol content, %.

The formula for biodiesel yield (Eq. 2.2) is given as:

$$Y(\%) = \frac{m_1 \times w}{m_2} \times 100 \tag{2.2}$$

m_1: mass of crude glycerol, g; w: glycerol content, %; m_2: theoretical mass of crude glycerol, g; Y: biodiesel yield, %.

2.1.3 Catalyst characterization

2.1.3.1 ICP-MS analysis

ICP-MS analysis showed that the main metal elemental composition of cedar ash was 16.47% Ca, 1.68% Mg, 2.39% Al, 0.20% Na, and 1.59% K. The Ca content of cedar ash is 16.47%, which is close to half the Ca content of ordinary limestone [3]; thus cedar ash has potential for use as a calcium-based solid base catalyst. In addition, both cedar ash and camphor tree ash contain a small amount of Mg, Al, Na, K, and other elements.

2.1.3.2 TG analysis

Fig. 2.1 shows the TG curve for cedar ash. It can be seen that the decomposition of cedar ash can be roughly divided into two steps. The weight loss is 25.30% over the whole process. In the first step, the weight loss at 37–200°C is 1.26% due to the removal of wetting water and capillary water in the cedar ash [4]. In the second step, the weight loss at 600–800°C is 24.04%. As seen in Fig. 2.2, the thermal decomposition of $CaCO_3$ results in the formation of CaO and CO_2 [5]. Fig. 2.1 also shows the TG curve of the precursor of cedar ash catalyst (calcination at 800°C, then hydration

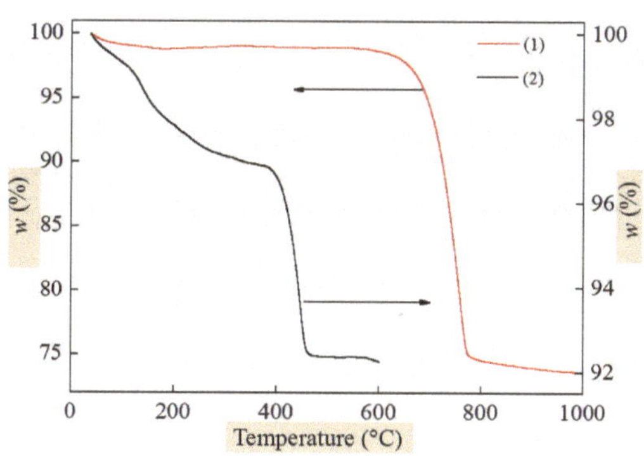

Fig. 2.1 TG curves of raw materials and the precursor of cedar ash catalyst: (1) TG curve of cedar ash and (2) TG curve of the precursor of cedar ash catalyst.

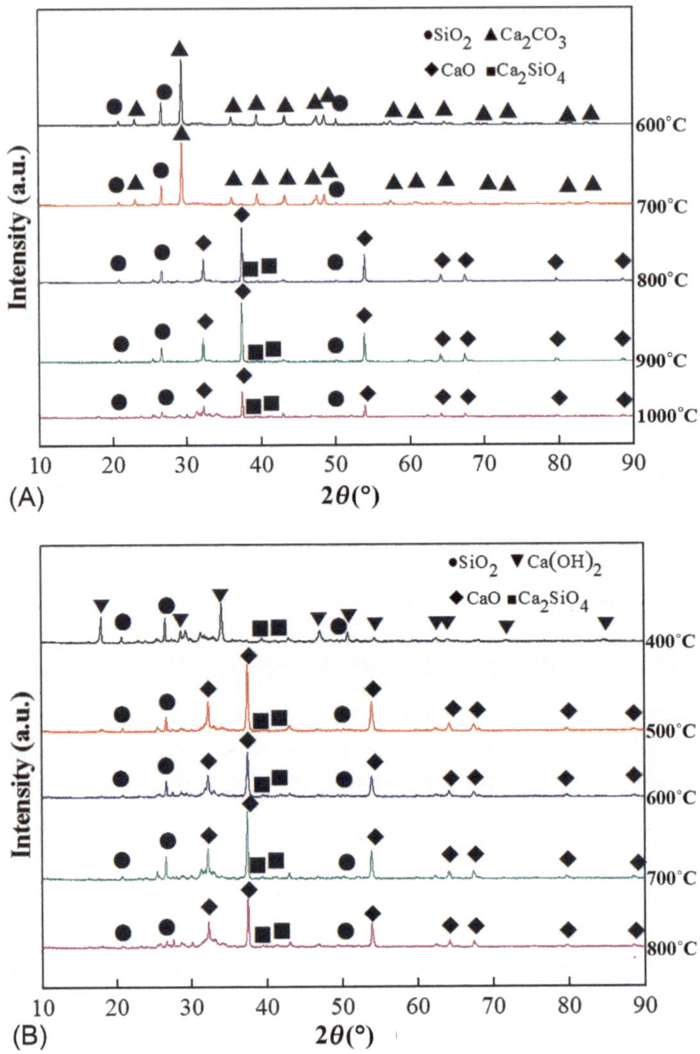

Fig. 2.2 XRD patterns of cedar ash by first and second calcining in the different temperatures. (A) XRD patterns of cedar ash by first calcining in the range of 600–1000°C and (B) XRD patterns of the precursor of cedar ash catalyst by second calcining in the range of 400–800°C.

and drying process). The decomposition of the cedar ash catalyst precursor can be roughly divided into three steps. The weight loss is 7.68% over the whole process. In the first step, the weight loss is 2.23% at 36–200°C due to the removal of wetting water and capillary water of the cedar ash precursor. In the second step, the weight loss is 0.90% at 200–360°C, which results in the thermal decomposition of other inorganic chemical combined water. In the third step, the weight loss is 4.55% at 360–470°C. Fig. 2.2B shows the mainly thermal decomposition of $Ca(OH)_2$, resulting in the formation of CaO and H_2O [6].

2.1.3.3 XRD analysis

Fig. 2.2A shows XRD patterns of cedar ash from the first calcining at 600–1000°C. It can be seen in the figure that the cedar ash presents a series of different diffraction peaks at different temperatures with the diffraction angle. As the cedar ash was calcined at 600 and 700°C, respectively, its angle 2θ = 23.1 degrees, 29.4 degrees, 36.0 degrees, 39.4 degrees, 43.2 degrees, 47.1 degrees, 48.5 degrees, 57.4 degrees, 60.7 degrees, 64.7 degrees, 69.2 degrees, 72.9 degrees, 81.5 degrees, and 83.8 degrees, in which there was a strong diffraction peak, belonging to the characteristic diffraction peaks of $CaCO_3$. This shows that $CaCO_3$ in cedar ash could not form CaO and CO_2 when the calcination temperature was below 700°C.

In addition, the diffraction peaks of 2θ = 20.9 degrees, 26.6 degrees, and 50.1 degrees belong to the typical diffraction peaks of SiO_2, which indicates that the ash contains silicon [4]. As cedar ash was calcined at 800–1000°C, its 2θ = 32.2 degrees, 37.4 degrees, 53.9 degrees, 64.2 degrees, 67.4 degrees, 79.7 degrees, and 88.5 degrees with a strong diffraction peak, which characterized the diffraction peaks of CaO. At the same time, it was found that the typical diffraction peaks of $CaCO_3$ disappeared. This indicates that when the calcination was carried out at 800–1000°C, $CaCO_3$ in the cedar ash was completely decomposed into CaO and CO_2. In addition, as the calcination was done at 800–1000°C, it was found that the diffraction peaks of new material 2θ = 39.5 degrees and 41.2 degrees, the same as the typical diffraction peaks of Ca_2SiO_4; and the diffraction peaks with 2θ = 20.9 degrees, 26.6 degrees, and 50.1 degrees were the typical diffraction peaks of SiO_2. This indicates that there was SiO_2 as the calcination took place at 800–1000°C, but the partial SiO_2 reacted with CaO, resulting in the formation of Ca_2SiO_4. This is similar to the results of Sharma's study [4].

Fig. 2.2B shows XRD patterns of the precursor of cedar ash catalyst (calcination at 800°C, then hydration and drying process) by second calcining at 400–800°C. Fig. 2.2B shows that the precursor of cedar ash catalyst at the second calcination temperature of 400°C, at 2θ = 18.0 degrees, 28.7 degrees, 34.1 degrees, 47.1 degrees, 50.8 degrees, 54.4 degrees, 62.6 degrees, 64.2 degrees, 71.8 degrees, and 84.8 degrees, had a strong diffraction peak belonging to the typical diffraction peaks of $Ca(OH)_2$. This indicates that $Ca(OH)_2$

in the precursor of the cedar ash catalyst could not be decomposed at the second calcination temperature of 400°C. For the precursor of cedar ash catalyst at the second calcination temperature range of 500–800°C, at 2θ = 32.2 degrees, 37.4 degrees, 53.9 degrees, 64.2 degrees, 67.4 degrees, 79.7 degrees, and 88.5 degrees, there was a strong diffraction peak belonging to the characteristic diffraction peaks of CaO. At the same time, the typical diffraction peaks of $Ca(OH)_2$ disappeared. This indicates that when the second calcination temperature is higher than 500°C, $Ca(OH)_2$ in the cedar ash is completely decomposed into CaO and H_2O. In addition, at a second calcination temperature range of 400–800°C, the diffraction peaks of 2θ = 39.5 degrees and 41.2 degrees, belong to the typical diffraction peaks of Ca_2SiO_4. The diffraction peaks of 2θ = 20.9 degrees, 26.6 degrees, and 50.1 degrees belong to the typical diffraction peaks of SiO_2. This indicates that Ca_2SiO_4 and SiO_2 still exist after the second calcination.

2.1.3.4 SEM analysis

Fig. 2.3A shows an SEM image of cedar ash with a lax and irregular porous structure. Fig. 2.3B shows an SEM image of cedar ash after the first calcining at 800°C. Combined with Fig. 2.2A analysis, the $CaCO_3$ of cedar ash has been completely converted to CaO, the crystal particles of which are mainly small aggregated blocks [7], and the particle surface is smooth. A typical diameter of crystal particles of cedar ash after first calcining is about 0.5 μm. Fig. 2.3C shows an SEM image of a cedar ash catalyst (second calcining at 500°C). Combined with Fig. 2.2B analysis, $Ca(OH)_2$ in the precursor of cedar ash was completely decomposed into CaO and H_2O. Compared with the first calcination, its crystal particles still have no regular shape, but the second calcination significantly reduced crystal particle size. A typical size of a crystal particle of cedar ash catalyst is about 0.1 μm. This indicates that the second calcination can greatly improve the particle size of the catalyst particles and increase the effective contact area of the catalyst, which is the same conclusion as that of Yung [8].

2.1.3.5 Base strength of catalyst

The Hammett indicator method can be used to determine the base strength of a catalyst. Phenolphthalein (*H*- = 9.8), 2,4-dinitroaniline (*H*- = 15.0), and 4-nitroaniline (*H*- = 18.4) were selected to measure the base strength. When the Hammett indicator color changes from an acidic to a basic color, it indicates that the base strength of the catalyst was greater than the *H*- value of the indicator. The results showed that cedar ash catalyst can cause

(A)

(B)

(C)

Fig. 2.3 SEM images of cedar ash by noncalcining, first and second calcining: (A) SEM image of cedar ash; (B) SEM image of cedar ash by first calcining at 800°C; (C) SEM image of cedar ash catalyst.

the phenolphthalein to turn red, but it cannot make 4-nitroaniline turn orange, nor 2,4-dinitroaniline turn to mauve. This indicates that cedar ash catalyst base strengths are $9.8 < H- < 15.0$, which is a typical value of a solid base catalyst [9].

2.1.4 Effect of catalyst preparation condition on biodiesel yield

When evaluating the activity of the cedar ash catalyst, the transesterification reaction conditions were considered as follows: cedar ash catalyst concentration of 6.0 wt%, methanol/oil molar ratio of 14:1, reaction time 5 h, and reaction temperature of 65°C. By considering biodiesel yield as a measure of the standard of catalytic activity, the effects of first calcination temperature, first calcination time, second calcination temperature, and second calcination time on the biodiesel yield were investigated. The results are shown in Fig. 2.4A–D.

In Fig. 2.4A the biodiesel yield increases significantly with calcination temperature at the first calcination temperature below 800°C. The biodiesel yield reaches 90.29% at the first calcination temperature of 800°C, and it gradually decreases with increasing calcination temperature. It was found that there is an optimum calcination temperature: a first calcination temperature that is too high or too low will significantly reduce catalyst activity. Combined with Fig. 2.2A, this shows that $CaCO_3$ in cedar ash does not cause thermal decompose when the first calcination temperature is too low. Once the first calcination temperature reaches 800°C, $CaCO_3$ in cedar ash has been completely converted to CaO. However, a first calcination temperature that is too high will promote the reaction of CaO and SiO_2, resulting in generation of Ca_2SiO_4. The content of the active ingredient CaO is thus reduced and the alkaline active sites of the cedar ash catalyst are decreased, which causes biodiesel yield to be decreased [4].

As shown in Fig. 2.4B, as the first calcination time increases from 1 to 2 h, the biodiesel yield increases. When the first calcination time is 2 h, the biodiesel yield reaches 90.37%. The yield is not significantly increased with further increase in the first calcination time. When the first calcination time reaches 5 h, the biodiesel yield is only 89.81%. This shows that the $CaCO_3$ in cedar ash has been completely converted to CaO during the first calcination in a relatively short period. Thereby CaO is further hydrated and changed by the cedar ash catalyst after the second calcination.

In Fig. 2.4C, it can be seen that as the second calcination temperature increases from 400°C to 500°C, the biodiesel yield increases significantly. When the second calcination temperature is 500°C, the biodiesel yield

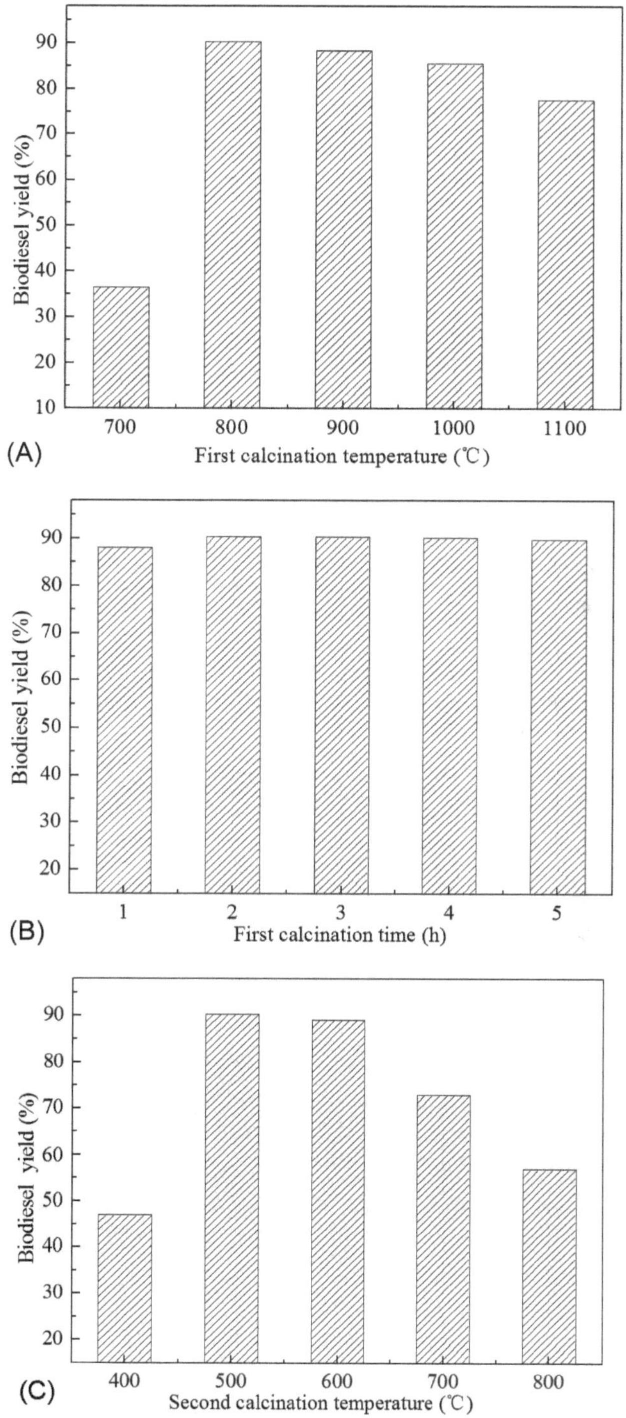

Fig. 2.4 See figure legend on next page.

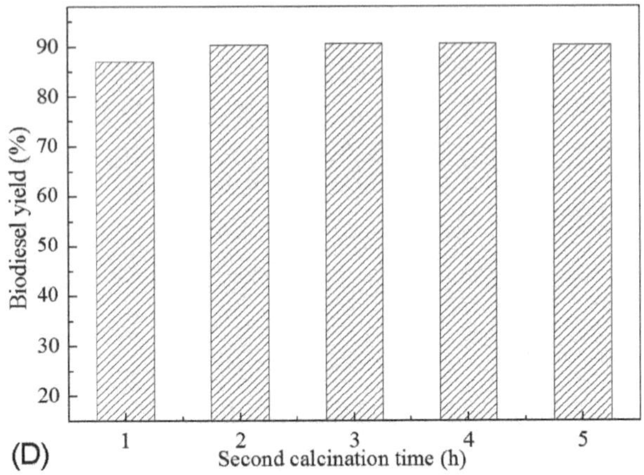

Fig. 2.4 Effect of catalyst preparation conditions on biodiesel yield. (A) First calcination temperature (first calcination time = 3 h, second calcination temperature = 500°C, second calcination time = 2 h); (B) first calcination time (first calcination temperature = 800°C, second calcination temperature = 500°C, second calcination time = 2 h); (C) second calcination temperature (first calcination temperature = 800°C, first calcination time = 2 h, second calcination time = 2 h); and (D) second calcination time (first calcination temperature = 800°C, first calcination time = 2 h, second calcination temperature = 500°C).

reaches 90.37%; but the yield is significantly reduced with further increasing the calcination temperature. This indicates that there is also an optimum second calcination temperature. A second calcination temperature that is too high or too low also affects the activity of the catalyst. Combined with Fig. 2.2B, this shows that the active component of cedar ash is still in an inactive state when the second calcination temperature is too low, indicating that $Ca(OH)_2$ is not decomposed into CaO. As the second calcination temperature is 500°C, the active component of cedar ash changes from the inactive state to the active state, meaning that $Ca(OH)_2$ is thermally decomposed to CaO. When the second calcination temperature is too high, it promotes CaO particle agglomeration, which reduces the surface area and catalytic activity of the catalyst [10], and causes biodiesel yields to decrease.

In Fig. 2.4D, it can be seen that as the second calcination time increases from 1 to 2 h, the biodiesel yield gradually increases. When the second calcination time is 2 h, the biodiesel yield reaches 90.37%. The yield increases only slightly with longer times. When the second calcination time is 3 h, the biodiesel yield reaches 90.71%. Based on energy savings, the second calcination time is chosen as 2 h, indicating that the active component of the catalyst can change from an inactive to an active state in a short second calcination time. Thus the catalyst has a basic active site.

In summary, the optimum conditions for the preparation of cedar ash catalyst are as follows: first calcination temperature of 800°C, first calcination time 2 h, second calcination temperature of 500°C, and second calcination time 2 h.

2.1.5 Effect of transesterification reaction conditions on biodiesel yield

Cedar ash catalyst was prepared under the preceding optimal conditions and used for transesterification to synthesize biodiesel. The effects of catalyst concentration, reaction time, methanol/oil molar ratio, and reaction temperature on the biodiesel yield were also investigated. The results are shown in Fig. 2.5A–D.

In Fig. 2.5A, when the catalyst concentration is less than 7.0 wt%, the biodiesel yield gradually increases with catalyst concentration. When the catalyst concentration is 7.0 wt%, the biodiesel yield reaches 91.52%; but after that point the biodiesel yield decreases with catalyst concentration. This shows that when catalyst concentration is too low, the concentration of the reaction system basic active sites is also low. In the same reaction time, the reaction rate is lower and biodiesel yield is lower as well. But when the catalyst concentration is more than 7.0 wt%, there is excessive catalyst in the reaction system, which increases the mass transfer resistance of the system and is not conducive to full contact between the reactants and the catalyst [11].

In Fig. 2.5B, when the reaction time is less than 5 h, the biodiesel yield increases gradually with the reaction time. When the reaction time is 5 h, the biodiesel yield reaches 91.52%. However, the biodiesel yield does not increase significantly with further increase in the reaction time. This shows that the biodiesel yield significantly increases with the reaction time and the reaction rate becomes faster at the early stage of transesterification. When the reaction time exceeds 4 h, the biodiesel yield increases slowly. This is because the concentrations of methanol and soybean oil in the reaction system are reduced, the biodiesel and glycerol concentrations are increased, and the reaction rate decreases at the late stage of the transesterification reaction. When the reaction time exceeds 5 h, the biodiesel yield stays constant, largely because the transesterification is approaching equilibrium [12].

In Fig. 2.5C, when the methanol/oil molar ratio is less than 14:1, the biodiesel yield gradually increases with the methanol/oil molar ratio. When the methanol/oil molar ratio is 14:1, the biodiesel yield reaches 91.52%; but the biodiesel yield then decreases with the methanol/oil molar ratio. This shows that if the methanol/oil molar ratio is relatively low, excess methanol can promote the transesterification reaction and improve biodiesel yield.

However, when the methanol/oil molar ratio exceeds 14:1, the reaction system has a higher concentration of methanol, which increases the glycerol solubility, which leads to an increase in glycerol concentration in the reaction system. Thereby, it prevents the transesterification, which is similar to Kotwal's findings [13].

In Fig. 2.5D, when the reaction temperature is below 65°C, the biodiesel yield gradually increases with the reaction temperature. When the reaction temperature is at 65°C, biodiesel yield reaches 91.52%; but the biodiesel yield then decreases with an increase in the reaction temperature. This shows that there is an optimum reaction temperature. When the reaction temperature is too low, it may reduce the three-phase (methanol-soybean oil-cedar ash catalyst) viscosity of the reaction system with temperature increase, which improves the compatibility of methanol and soybean oil, and promotes mutual diffusion of the soybean oil and methanol, and thereby increases the reaction rate. However, when the reaction temperature exceeds 65°C, the reaction temperature of the system exceeds the boiling point of methanol. Methanol is converted from the liquid phase to the gas phase, leading to a decrease in the methanol concentration in the reaction system. This is bad for transesterification, as Birla also found [14].

In conclusion, the best conditions for synthesis of biodiesel catalyzed by cedar ash catalyst are as follows: catalyst concentration 7.0 wt%, reaction time 5 h, methanol/oil molar ratio 14:1, and reaction temperature 65°C.

2.1.6 Reusability of the catalysts

The cedar ash catalyst was prepared under optimal conditions. Using this catalyst, the transesterification was also conducted under the best biodiesel synthesis conditions. The catalyst was recovered between each reaction by filtration, thoroughly washed with methanol, and then used again [15]. The result is shown in Fig. 2.6. From Fig. 2.6A, it can be seen that in the range of four times of repeated use, the cedar ash catalyst still has a stronger diffraction peak at $2\theta = 32.2$ degrees, 37.4 degrees, 53.9 degrees, 64.2 degrees, 67.4 degrees, 79.7 degrees, and 88.5 degrees, belonging to the characteristic diffraction peaks of CaO. The diffraction peaks of $2\theta = 20.9$ degrees, 26.6 degrees, and 50.1 degrees are typical diffraction peaks of SiO_2. In addition, the diffraction peaks of $2\theta = 39.5$ degrees and 41.2 degrees are typical diffraction peaks of Ca_2SiO_4. In the repeated use (two to four times), the range of diffraction peak $2\theta = 18.0$ degrees and 34.1 degrees, belonging to $Ca(OH)_2$. This indicates that the active component in the cedar ash catalyst is still CaO after four times of repeated use, but a portion of CaO has been

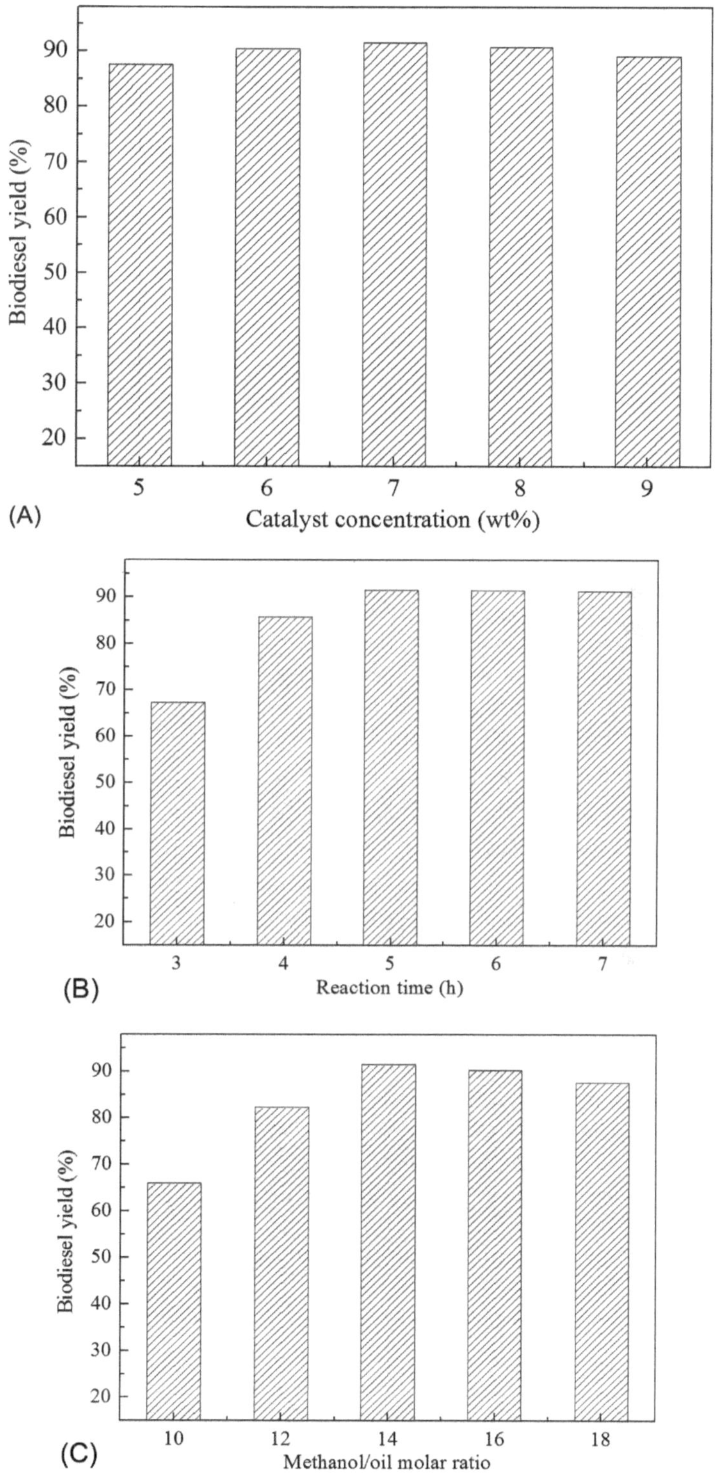

Fig. 2.5 See figure legend on next page.

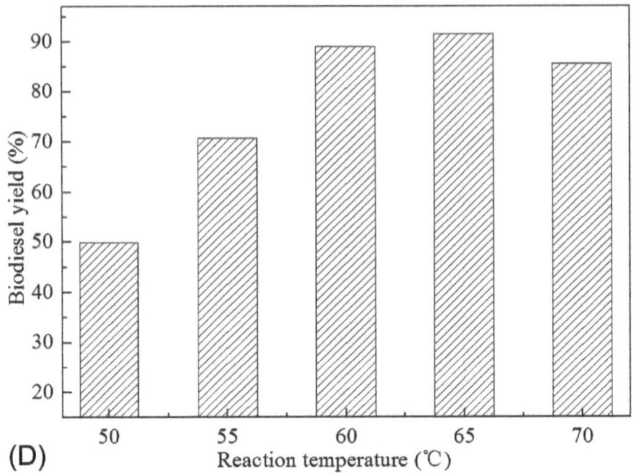

Fig. 2.5 Effect of transesterification reaction conditions on biodiesel yield. (A) Catalyst concentration (reaction time = 5 h, methanol/oil molar ratio = 14:1, reaction temperature = 65°C). (B) Reaction time (catalyst concentration = 7%, methanol/oil molar ratio = 14:1, reaction temperature = 65°C). (C) Methanol/oil molar ratio (reaction time=5 h, catalyst concentration = 7%, reaction temperature = 65°C). (D) Reaction temperature (reaction time = 5 h, catalyst concentration = 7%, methanol/oil molar ratio = 14:1).

converted to $Ca(OH)_2$ during the two to four times of reuse. This finding is similar to the results of Maneerung's study [16]. As shown in Fig. 2.6B, the catalytic activity of the catalyst was obviously decreased after several times of reuse. With the fourth time, the biodiesel yield was reduced to 82.03%. When combined with the Fig. 2.6A analysis, it was found that CaO was partially converted to $Ca(OH)_2$. This is because it is easy for CaO to react with water from air and the reaction system, resulting in the reduction of the catalyst activity center [16].

The chemical equations of cedar ash catalyst for the two calcination processes can be summarized as follows:

First calcination (800°C, 2 h):

$$CaCO_3 \rightarrow CaO + CO_2$$

$$2CaO + SiO_2 \rightarrow Ca_2SiO_4$$

Second calcination (500°C, 2 h):

$$Ca(OH)_2 \rightarrow CaO + H_2O$$

Through these chemical equations, it can be found that the active species of cedar ash catalyst was CaO.

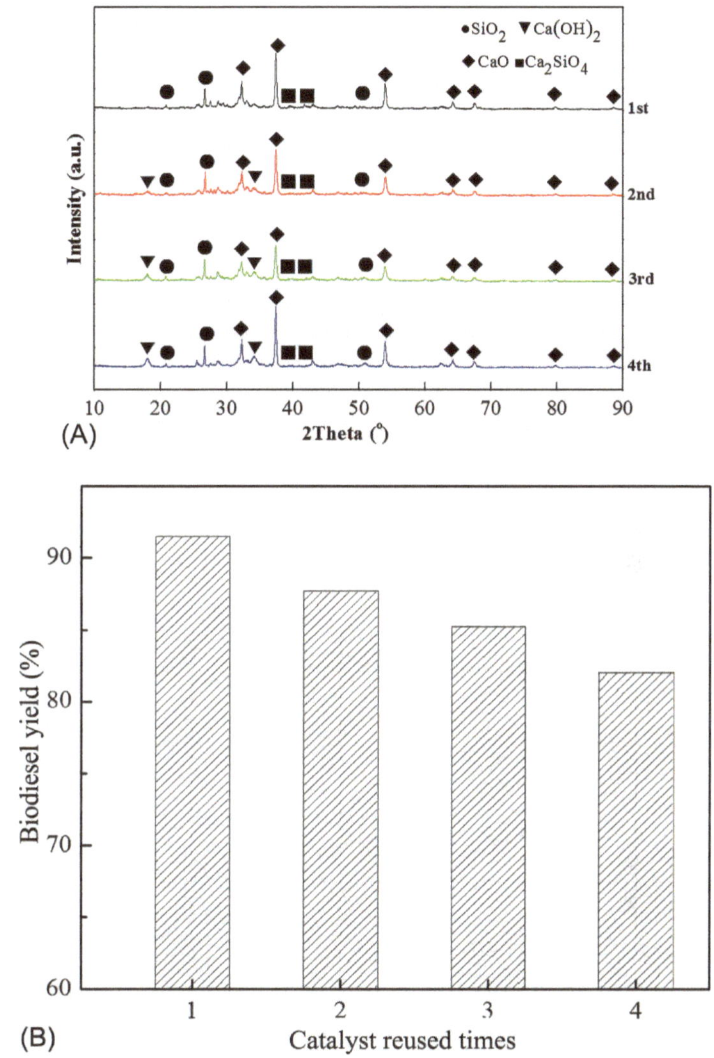

Fig. 2.6 Reusability of cedar ash catalyst: (A) XRD patterns of the used cedar ash catalysts after each reaction cycle and (B) effect of the reusability of cedar ash catalyst on biodiesel yield.

2.2 Characterization of biodiesel soot particles

2.2.1 Introduction

To alleviate global climate change, reduce dependency on fossil fuels, and improve energy security, scientists in various countries are developing new clean fuels, among which biodiesel is a very promising alternative [17]. Biodiesel is the product of the transesterification of animal oil and vegetable oil. The main component of biodiesel is fatty acid methyl ester. Biodiesel has excellent environmental protection characteristics and biodegradability, and it is not volatile. It can be used as a fuel alone or mixed with petrochemical diesel in any proportion [18]. Therefore, countries around the world are taking measures to promote the use of biodiesel.

During engine operation, soot particles are produced during fuel combustion. Some soot particles enter the atmosphere via exhaust emission systems, which aggravates environmental pollution, while other soot particles enter the crankcase through the piston ring, thus polluting the lubricating oil. As soot particles are charged, they tend to agglomerate, increasing the viscosity of the lubricating oil, a typical problem faced in diesel engines [19]. In fact, there are obvious differences in the morphology, composition, and structure of soot particles produced by different fuels [20], which can have a great impact on their agglomeration characteristics in lubricating oil [21]. Zhang Bin et al. [22, 23] compared biomass fuel soot with conventional diesel soot and found that the oxygen content in biomass fuel soot was higher than that of conventional diesel fuel, and the graphitization degree was also higher. Clague et al. [24] found that there were significant differences in the element composition and surface chemical functional groups between diesel soot and soot stimulant—carbon black. Wang et al. [25] investigated the soot generated by the combustion of biodiesel prepared from different raw materials and pointed out that different saturation and oxygen content in biodiesel resulted in different amounts of soot formation. Savic et al. [26] studied the soot produced by the combustion of biodiesel and petrochemical diesel in different proportions and found that the primary particle size of the soot decreased with an increase of the biodiesel content. Sharma et al. [27] analyzed the structure and chemical composition of diesel soot extracted from crankcase lubricating oil and soot emitted from exhaust pipes using various characterization techniques. It was found that the main elements of the two soot samples were C and O, but the diesel soot extracted from the crankcase lubricating oil also contained Ca, Zn, P, S, and other elements. Zeidan et al. [28] found that an excessive increase in the viscosity of lubricating oil was closely related to the formation of large soot aggregates in the oil.

As the main component of biodiesel is fatty acid methyl ester, while the main components of petrochemical diesel are alkanes and aromatic hydrocarbons, there will inevitably be great differences in the composition, morphology, and structure of soot generated by the combustion of these two fuels [29]. Therefore, further study is necessary. Based on a comparison of the morphology, composition, and structure of biodiesel soot (BDS) and diesel soot (DS), this chapter analyzes the effect of the particle characteristics of the two kinds of soot particles on their agglomeration characteristics in liquid paraffin (LP, a lubricant base oil simulant). This study aims to lay a foundation for solving the dispersion problem of BDS in lubricating oil and to make biodiesel more widely used in engines by providing a test basis.

2.2.2 Materials and methods

2.2.2.1 Experimental materials

The biodiesel used here was produced from methanol and soybean oil, using a self-made cedar ash catalyst, through transesterification [30]. The $0^{\#}$ diesel was purchased from China Petroleum and Chemical Corporation. According to National Standards, the physical and chemical properties of the two kinds of fuels were determined using standard laboratory oil performance testing equipment. The relevant parameters are shown in Table 2.1. The LP was purchased from Hengshui Diyi Petrochemical Co., Ltd.

2.2.2.2 BDS and DS preparation

BDS and DS were prepared as follows: a certain amount of biodiesel was poured into a 100 mL porcelain crucible, and the crucible was placed on an iron platform and a piece of clean glass plate (i.e., a soot capture board) was placed above it. The center of the glass plate was located 2 cm above a flame to obtain efficient soot capture (NB: during initial tests, it was found that the amount of soot produced increased by 25% as the distance of the flame changed from 10 to 2 cm; hence, a distance of 2 cm is used here), under normal temperature and atmospheric pressure. After the biodiesel was burned, a medication spoon was used to scrape off the soot deposited on

Table 2.1 Physical and chemical properties of biodiesel and No. 0 diesel oil.

Items	No. 0 diesel	Biodiesel	Methods
Density (kg/m^3)	815	875	GB/T 1884-2000
Kinematic viscosity (mm^2/s)	2.51	4.49	GB/T 265-1988
Acid value (mg/g)	0.11	0.045	GB/T 264-1983
Flash point (closed up) (°C)	55	168	GB/T 261-2008
Mechanical admixtures (%)	–	–	GB/T 511-2010
Solidification point (°C)	–6	–12	GB/T 510-1983
Mass fraction of the sulfur (%)	0.04	0	GB/T 380-1977
Copper strip corrosion (50°C, grade)	1	1	GB/T 5096-1985

the glass plate, which was placed in a vacuum and oven dried at 110°C for 3 h. Finally, the BDS was obtained by grinding it into a uniform powder in an agate mortar. DS was produced by burning $0^{\#}$ diesel oil in the same way.

The combustion of 1000 g of $0^{\#}$ diesel yielded 25–32 g of soot, and that of 1000 g of biodiesel yielded 13–18 g of soot. This indicates that after burning these two fuels, which had the same initial masses, the amount of soot produced by biodiesel combustion was less than that produced by $0^{\#}$ diesel. This is similar to the results of Kohse-Höinghaus [31], who found that biodiesel had higher oxygen content than petrochemical diesel, which helped to reduce the generation of soot precursors, resulting in a smaller amount of soot being produced by biodiesel.

2.2.2.3 Analysis of the morphology, structure, and composition of the soot particles

The flame temperature was measured using a data collector (Agilent 34970A) and a K-type thermocouple (Ni, Cr-Ni, and Si, temperature range 0–1300°C). Field emission transmission electron microscopy (FETEM, JEM-2100F) was used to analyze the morphology and structure of the primary particles of the BDS and DS. X-ray diffraction (XRD, D/MAX2500V) was used to analyze the main phases of the soot particles. The disorder degrees of the BDS and DS were analyzed by Raman spectroscopy (HR Evolution). The functional groups on the surfaces of the two fuels and their soot particles were analyzed by Fourier transform infrared spectroscopy (FTIR, Nicolet6700). X-ray photoelectron spectroscopy (XPS, ESCALAB250) was used to analyze the element content and chemical state of the surface of BDS and DS. The element content in the soot particles was detected by an elemental analyzer (Vario EL cube). The particle sizes of the BDS and DS in the LP were measured by a zeta potentiometer (Nano-ZS90).

2.3 Physical-chemical properties of biodiesel soot

2.3.1 BDS and DS surface morphology analysis

Fig. 2.7 shows an FETEM image of the BDS and DS. The image-processing software Nano Measurer was used to check 10 particles to calculate the average diameter of the soot's primary particles. The average diameter of the primary particles of BDS was 35 nm and that of the primary particles of DS was 39 nm, as shown in Fig. 2.1A and B, respectively. The smaller diameter of the primary particles of BDS is due to the biodiesel containing oxygen, which promotes combustion. Both the BDS and DS showed aggregated

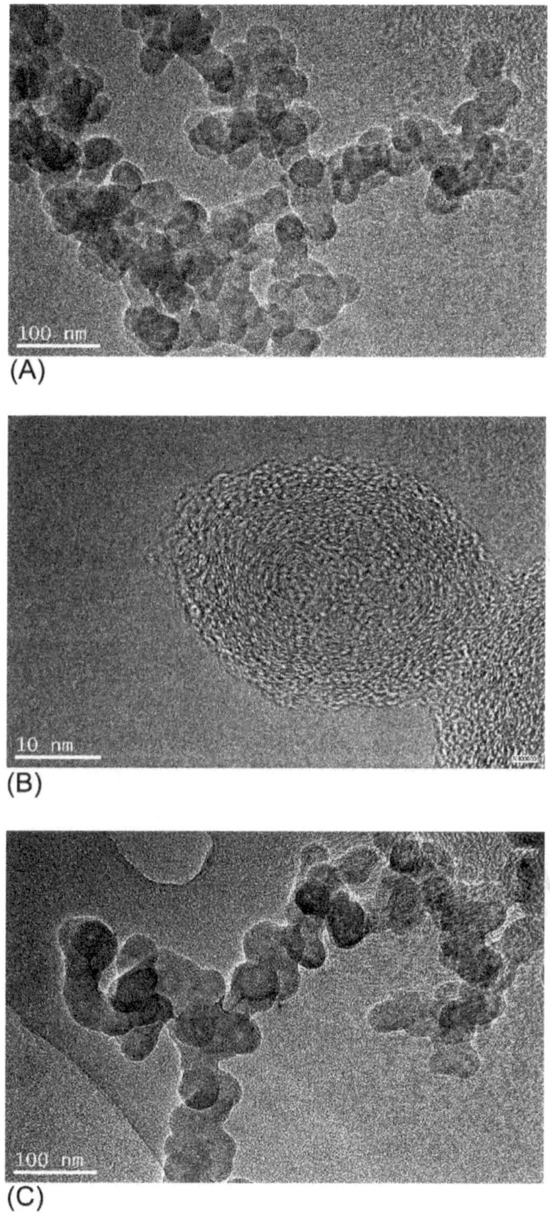

Fig. 2.7 See figure legend on next page.

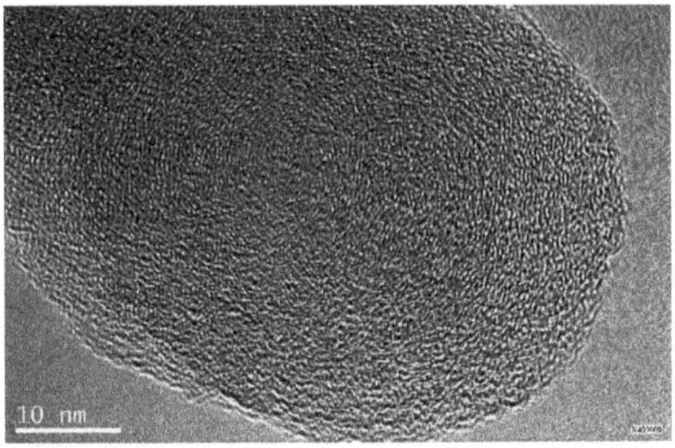

(D)

Fig. 2.7 FETEM images of BDS and DS: (A) FETEM image of BDS (100 nm), (B) FETEM image of BDS (10 nm), (C) FETEM image of DS (100 nm), and (D) FETEM image of DS (10 nm).

chains composed of a large number of nearly spherical particles. The flame temperature of 0# diesel was 590.3°C, and that of biodiesel was 625.9°C. The latter's larger flame temperature promoted more chemical reactions and led to the primary particle diameter of the BDS being smaller [20].

Fig. 2.7C shows that there were a large number of discontinuous and concentric oriented graphite fragments in the outer layer of the BDS, while the inner core had an amorphous carbon structure. Fig. 2.7D shows that the DS had an amorphous carbon structure; it also contained some discontinuous graphite fragments. This is similar to the findings of [22, 32, 33], who found that petrochemical diesel soot had only a small number of short-range ordered graphite layers and that the BDS had a graphitized core-shell structure. This indicates that the graphitization degrees of the BDS and DS particles are significantly different. Therefore, analyzing the internal structure of the BDS and DS, as shown in Figs. 2.2 and 2.3, is necessary.

2.3.2 Structural characteristics of BDS and DS

Fig. 2.8 shows XRD patterns of the BDS and DS. It can be seen that with a change of the diffraction angle, the BDS and DS both showed a series of strong diffraction peaks at 2θ = 24.50 degrees and 43.59 degrees, which are classified as Graphite-2H (002) and (101) crystal surfaces (2θ = 26.38 degrees, 44.39 degrees). It is worth mentioning that the diffraction peaks of both the BDS and DS shifted to lower angles. The reason for this is that the grain sizes of the BDS and DS were less than 100 nm; thus, the small-sized nanoparticles caused the diffraction angle to deviate to smaller angles [34].

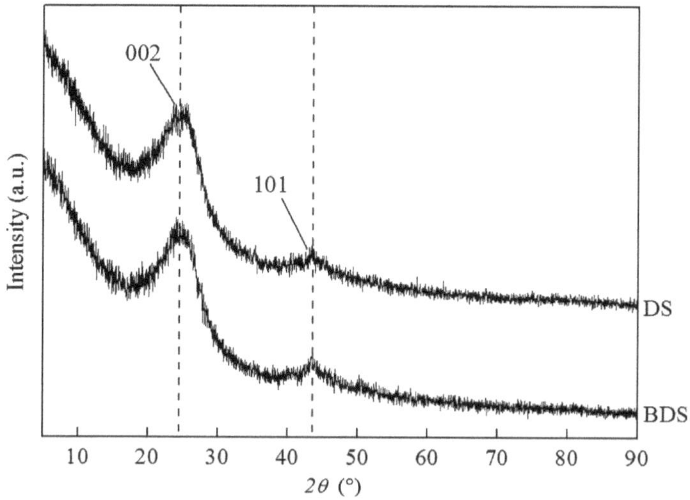

Fig. 2.8 XRD patterns of BDS and DS.

In addition, except for the diffraction peaks at 24.50 degrees and 43.59 degrees, the XRD patterns of the BDS and DS showed dispersed amorphous peaks, indicating that amorphous carbon and graphite existed in both the BDS and DS, which is consistent with the results of the FETEM analysis.

Fig. 2.9 shows Raman spectra of the BDS and DS, which were analyzed and displayed using the Origin 8.5 software. After Gaussian fitting, the ratio of peak area (D) to the peak area (G) for BDS and DS, respectively, was determined using the intensity ratio of two characteristic peaks (I_D/I_G). It can be seen that there are two characteristic peaks in the Raman spectra: Peak D (1280–1350 cm^{-1}), which is attributed to the disordered graphitic peak of the soot particles, and peak G (1580–1600 cm^{-1}), which is attributed to the graphitic peak of the soot particles [35].

The intensity ratio (I_D/I_G) of the two characteristic peaks reflects the degree of disorder of the soot particles. The larger the ratio, the higher the degree of disorder and hence the lower the degree of graphitization. Table 2.2 shows the Raman spectra parameters for the BDS and DS. It can be seen that the intensity ratio of DS (3.162) was greater than that of the BDS (2.937). This indicates that the DS had a higher degree of disorder, a larger number of crystal-structure defects, and higher amorphous carbon content. However, the graphitization degree of the BDS was higher, and the number of graphite fragments contained in the BDS was higher than that in the DS. This result agrees with the FETEM analysis results and is similar to the results of [36, 37], who found that the degree of disorder of DS is higher than that of BDS.

Table 2.2 Raman spectrum parameters of BDS and DS.

	D band		G band		
Soot samples	**Raman shift (cm)**	**FWHM**	**Raman shift (cm)**	**FWHM**	I_D/I_G
BDS	1338	288	1584	105	2.937
DS	1348	296	1580	103	3.162

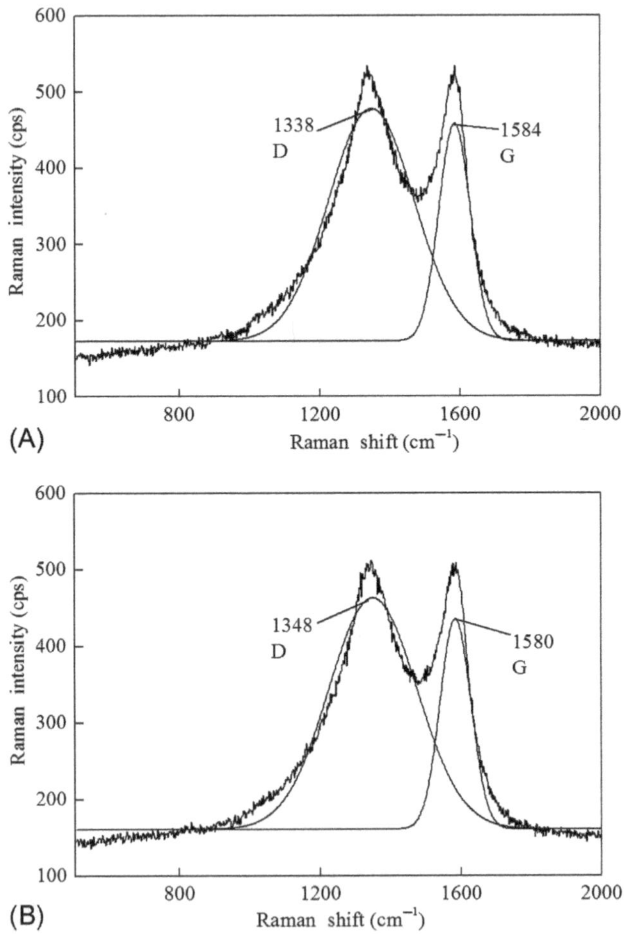

Fig. 2.9 Raman spectrum of BDS and DS:(A) BDS and (B) DS.

2.3.3 Chemical component of the BDS and DS

Fig. 2.10 shows FTIR spectra of the two kinds of fuel and their corresponding soot particles. As shown in Fig. 2.10A, the FTIR spectra of biodiesel and $0^{\#}$ diesel both contain saturated C-H stretching vibration peaks (2953, 2925, and 2853/cm) [38], C=C stretching vibration peaks due to olefin

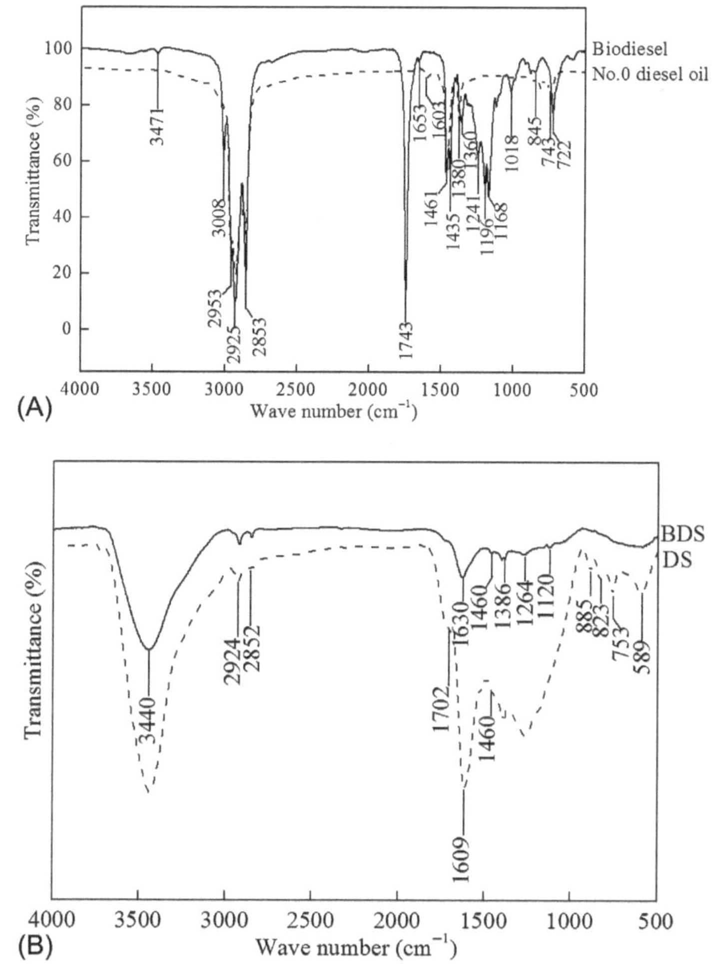

Fig. 2.10 FTIR spectra of diesel samples and their soots: (A) biodiesel and No. 0 diesel oil (B) BDS and DS.

and aromatic hydrocarbons (1653 and 1603/cm), flexural vibration peaks of —CH_2— and —CH_3 (1461 and 1435/cm), flexural vibration peaks of —CH_3 (1380 and 1360/cm^{-}), and peaks of —CH_2— (743 and 722/cm) [39]. Biodiesel showed some characteristic peaks not displayed by $0^{\#}$ diesel: the —C=O frequency doubling peak of the ester group (3471/cm), an unsaturated C—H stretch vibration peak (3008/cm), a —C=O stretching vibration peak of the ester group (1743/cm), a stretching vibration peak of C—O—C and C—C—O (1241/cm), the characteristic peaks of fatty acid methyl ester groups (1196 and 1168/cm) [38], and the flexural absorption vibration peaks of CH_3O— (1018 and 845/cm) [39]. These results indicate that, compared with $0^{\#}$ diesel, the particularity of biodiesel lies in its oxygen-containing functional groups (—C=O, CH_3O—, C—O—C, and C—C—O), where this additional oxygen facilitates combustion [20].

As shown in Fig. 2.10B, both the FT-IR spectra of the BDS and DS contain the —OH stretching vibration peak (3440/cm), saturated C—H stretching vibration peaks (2924 and 2852/cm), olefin C=C stretching vibration peaks (1630 and 1609/cm), deformation vibration peak of $—CH_3$ (1386/cm), asymmetric stretching vibration peak of the ether C—O—C (1264/cm) [40], the stretching vibration peak of C—O (1120/cm) and SO_4^{2-} (589/cm) [41], and the flexural vibration peak of N—H (1460/cm) [42].

It was found that the DS contained some characteristic peaks that were not displayed by the BDS, such as the deformation vibration peaks of aromatic hydrocarbon C—H (885, 823, and 753/cm) [40]. The absence of these peaks for the BDS is due to $0^{\#}$ diesel containing aromatic hydrocarbons. In addition, the DS also showed a —C=O stretching vibration peak (1702/cm) [42]. This shows that, as $0^{\#}$ diesel oil did not have oxygen-containing functional groups, its incomplete combustion produced —C=O functional groups in the organic surface residue produced by the DS. However, the biodiesel oil contains a large number of oxygen-containing functional groups, which promote combustion reactions. Biodiesel contained —C=O functional groups of organic matter that were burned into CO_2 or CO [20, 26], which resulted in the surface of the BDS not containing C=O functional groups. This is similar to the findings of [43], who found that petrochemical diesel soot contained oxygen-containing functional groups, such as —C=O and —C—O. However [22], thought that DS did not contain C—O—C functional groups. In their investigation, they used a different soot-capturing board and different flame distances (10 cm) than used in this work (2 cm). As the flame distance results in different captured soot particles and different surface oxidation degrees [33], the soot surface functional groups are different in our study and that of [43].

Fig. 2.11 shows XPS spectra of the BDS and DS obtained using the XPS PEAK software. As shown in Fig. 2.11A, the C_{1s} spectra of the BDS and DS contain C—C (284.8 eV) [23]. In addition, the C_{1s} spectra of the DS contain —C=O or C—O—C (287.9 eV), while the C_{1s} spectra of the BDS contain C—O—H or C—O—C (286.9 eV). In addition, the ratio of the peak area of the total lipophilic functional group (C—C) to the peak area of the total oleophobic functional group (—C=O, C—O—C, and C—O—H) indicates the lipophilicity of the material, where a larger λ implies stronger lipophilicity of the soot particles. It was found that $\lambda_{DS} = 3.45$ and $\lambda_{BDS} = 2.66$;

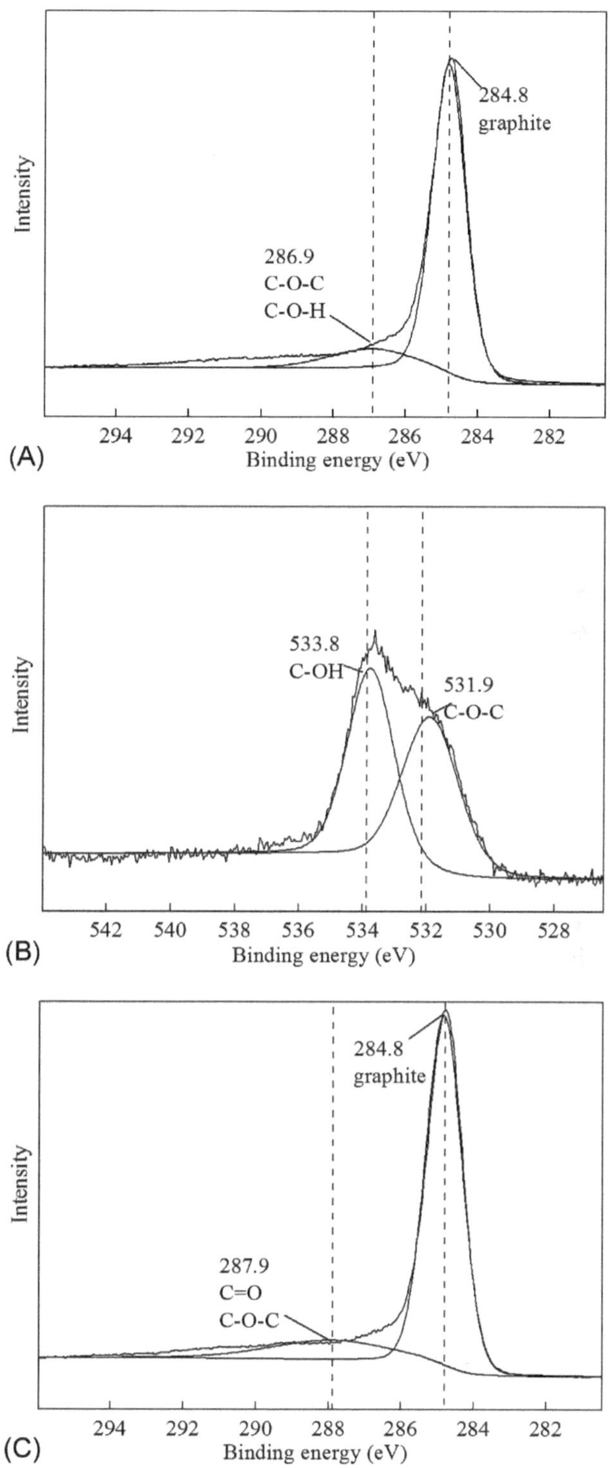

Fig. 2.11 See figure legend on next page.

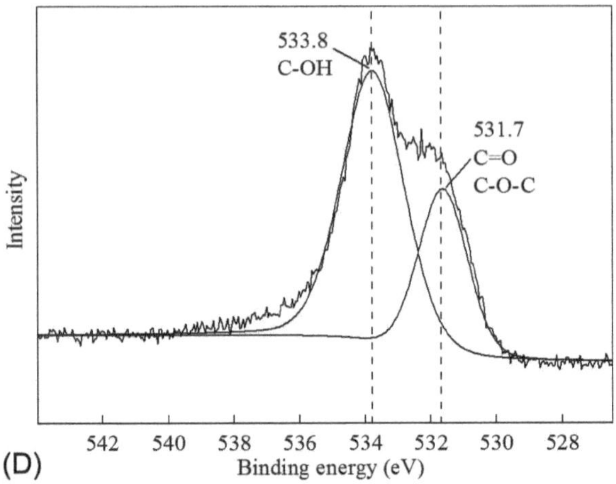

(D)

Fig. 2.11 XPS spectra of BDS and DS: (A) C_{1s} spectra of BDS, (B) O_{1s} spectra of BDS, (C) C_{1s} spectra of DS, and (D) O_{1s} spectra of DS.

that is, λ_{DS} was greater than λ_{BDS}. This indicates that the lipophilic property of the DS was stronger than that of the BDS. As shown in Fig. 2.11B, the C—OH bond (533.8 eV) was included in the O_{1s} spectra of both the BDS and DS, and in addition, the O_{1s} spectra of DS included —C=O or C—O—C (531.7 eV). This indicates that the two kinds of soot both contained C—C, C—O—C, and C—OH, but the BDS did not contain C=O. This result corresponds to the FTIR analysis results and is similar to the research results of [44], who found that the content of —C=O functional group in the soot of a biodiesel/petrochemical diesel mixture was lower than that of petrochemical diesel.

Table 2.3 shows the element compositions of the BDS and DS. It can be seen that both mainly contain C and O, while H, N, and S show relatively low concentrations. Among them, the S contained in the BDS and DS corresponded to the SO_4^{2-} characteristic peak seen in the FTIR spectra, which may be caused by the S contained in $0^{\#}$ diesel [29]. However, biodiesel contained almost no sulfur, and the sulfur in the BDS may have come from inorganic salts or residual contaminants imparted during the preparation

Table 2.3 Elemental composition of BDS and DS.

	XPS		Elemental analyzer				
Soot samples	**C (%)**	**O (%)**	**C (%)**	**O (%)**	**H (%)**	**N (%)**	**S (%)**
BDS	92.61	7.39	90.99	6.91	1.68	0.31	0.11
DS	91.48	8.52	88.24	8.60	2.30	0.57	0.29

of biodiesel. The H may have come from organic compounds produced by inadequate fuel combustion [23]. The N in the DS may have arisen due to the incomplete combustion of nitrogen-containing additives (such as amine antioxidants and low-temperature fluidity improvers) in the diesel. The N in the BDS may have been caused by the incomplete combustion of nitrogen-containing organic matter (such as amino acids) in the soybean oil in the feedstocks of biodiesel. According to the FTIR analysis results, the N shown in the DS and BDS spectra existed in the form of N—H functional groups. Compared with the DS, the BDS had a high content in C and a lower content in O, H, N, and S.

2.4 Aggregation properties of biodiesel soot

2.4.1 Characterization of BDS aggregation properties

2.4.1.1 Determination of the viscosity of LP containing soot particles

A rotational viscometer (NDJ-5S, Ningbo Weide Instrument Co., Ltd.) was used to measure the dynamic viscosity of the LP containing soot particles at 20°C with a No. 2 rotor operated at a speed of 30 rev/min. The effect of the soot particle content on the relative viscosity of the LP was subsequently investigated. The relative viscosity was calculated by the following formula (2.3):

$$\mu_r = \mu_d / \mu_0, \tag{2.3}$$

where μ_d is the dynamic viscosity (mPa·s) of the soot-containing LP, μ_0 is the dynamic viscosity of the LP (mPa·s), and μ_r is the relative viscosity (mPa·s) of the soot-containing LP.

2.4.1.2 BDS and DS contact angle measurements

We measured and weighed appropriate amounts of BDS and DS and placed them in a pressure plate mold. Next, we used a vertical hydraulic jack (QYL10T, Haiyan Guangyao machinery factory) to press them (pressure of 10 MPa) for 10 min. Then, we formed them into soot tablets with a diameter of 12 mm.

An optical contact angle/interface tensiometer (SL 200KS, Shanghai Solon Information Technology Co., Ltd.) was used to take pictures of the morphology of distilled water dripping on the soot tablet, and then the contact angle of the distilled water on the soot tablet was measured. The surface energy of the soot tablet was calculated by the following equation (Eq. 2.4) [45]:

$$\gamma_{sv} = \frac{\gamma_{lv}\left(\cos\theta + 1\right)^2}{4} \tag{2.4}$$

where γ_{sv} is the solid surface energy (mJ/m^2); and γ_{lv} is the liquid surface energy (mJ/m^2).

2.4.2 Studying the agglomeration characteristics

The viscosity change in a lubricating oil caused by soot pollution can directly reflect the degree of soot agglomeration therein [28]; therefore, viscosity can be used to study the agglomeration characteristics of soot. Fig. 2.12 shows the varying relative viscosity of different BDS and DS content in the LP. It can be seen that, with an increase of the mass fraction of BDS and DS, the relative viscosity of the lubricating oil showed an obvious upward trend. When the mass fraction of the soot was less than 1%, the relative viscosities of the LP containing two kinds of soot particles were similar. However, as the mass fraction of soot continued to increase, the relative viscosity of the LP containing the BDS became gradually higher than that containing DS under the same mass fraction of soot. The relative viscosity of the LP containing BDS was 8.852 mPa·s and that of LP containing DS was 7.991 mPa·s when the mass fraction of soot was 5%. The data in Fig. 2.12 were fitted with an exponential function, and the best-fitting relation for the DS was $\mu_r = 0.2909 + 1.0033e^{0.4059x}$, where x is the mass fraction of DS, and $R^2 = 0.9808$. For the BDS, the best-fitting relation

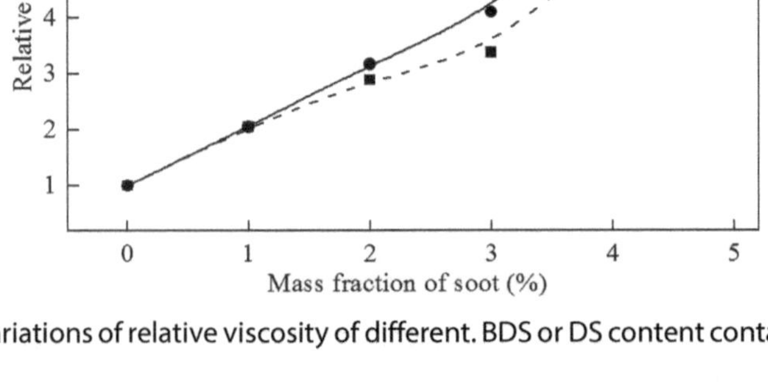

Fig. 2.12 Variations of relative viscosity of different. BDS or DS content contaminated LP.

was $\mu_r = -0.6340 + 1.8314e^{0.3271x}$, where x is the mass fraction of BDS, and $R^2 = 0.9921$. This indicates that the relative viscosity of LP increased exponentially with an increase of the mass fraction of BDS and DS. While both BDS and DS agglomerated in the LP [46], BDS had a greater influence on the viscosity of the LP than the DS.

2.4.3 Aggregate mechanism analysis

The optical contact angle/interfacial tensiometer was used to measure the contact angle of distilled water on the soot tablets; the results are shown in Fig. 2.13. The contact angles of the BDS and DS were 113 degrees and 127 degrees, respectively. The surface energy of a solid is related to its wetting properties: when the same liquid is used to measure the contact angle of different solid surfaces, the greater the surface energy of the solid, the easier the droplets spread on the solid's surface. We calculated the surface energy of

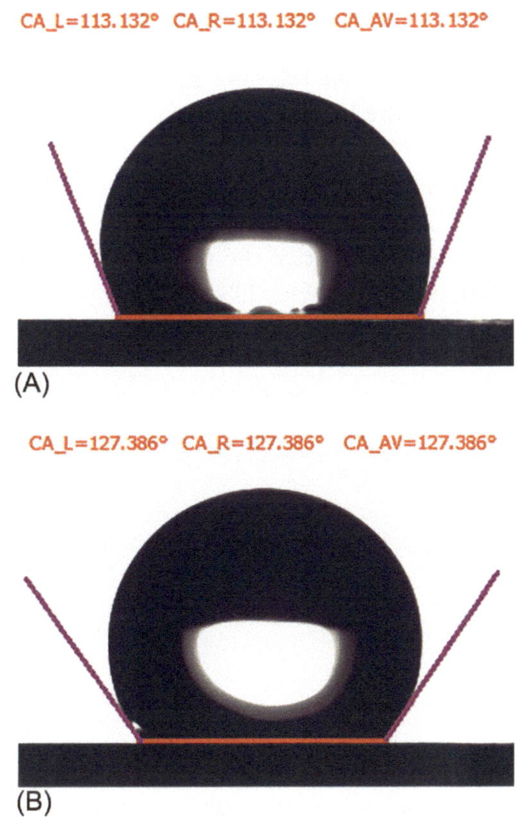

Fig. 2.13 Contact angle of BDS and DS: (A) BDS and (B) DS.

the BDS (γ_{sv}) to be 6.7 mJ/m^2, and the surface energy of the DS (γ_{sv}) was 2.9 mJ/m^2. For comparison, the surface energy of water in air (γ_{lv}) is 72.7 mJ/m^2. These results indicate that water is more easily adsorbed on the surface of the BDS and that oil is more easily adsorbed on the surface of the DS. This may be because the nonpolar groups on the surface of the DS can better mutually dissolve with the nonpolar groups in the LP (i.e., the XPS analysis shows that DS is more lipophilic than BDS). Compared with the DS, the BDS is more hydrophilic. The hydrophilic polar groups on the surface of the BDS and the nonpolar groups on the surface of the LP are mutually exclusive and BDS tended to aggregate into larger particles, which minimized contact with LP [21]; in other words, BDS aggregated more easily than the DS in the LP.

Using ultrasonic dispersion and the gradient dilution method, the amount of BDS and DS particle suspension, with mass fractions of 1.0×10^{-5}, in the LP was obtained. The particle size distributions of the agglomerated BDS and DS in LP were measured by the zeta potentiometer, as shown in Fig. 2.14. It can be seen that the particle size distribution range of the agglomerated BDS in the LP was 122–396 nm, while that of the agglomerated DS in the LP was 122–342 nm. The average particle size of the agglomerated BDS in the LP was 227.8 nm, while that of the agglomerated DS in the LP was 194.5 nm. The particle size distribution and average particle size of the agglomerated BDS were both larger than those of the agglomerated DS. Recall that the average diameter of the primary particles of the BDS was smaller than that of the primary particles of the DS. The reason for this is that, compared with DS, BDS has lower lipophilicity (the

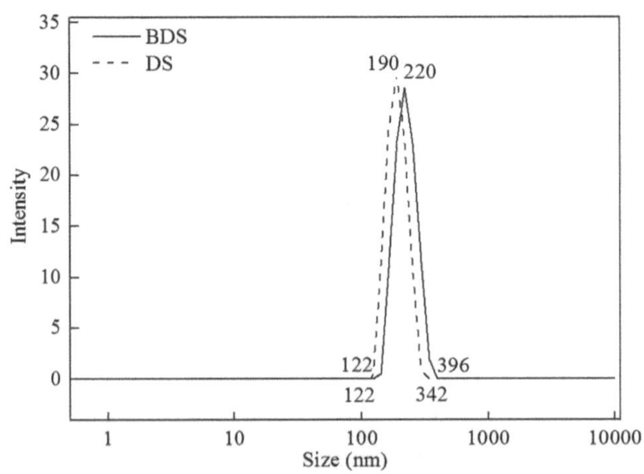

Fig. 2.14 Particle size distribution of BDS and DS in LP.

XPS and contact angle analyses confirmed this result), which means BDS more easily aggregated in the LP. This is why the BDS had a greater effect on the viscosity of LP than the DS.

2.5 Conclusions

The conclusions of this study are summarized in the following paragraphs.

Cedar ash catalyst can catalyze a transesterification reaction to synthesize biodiesel; the active species of cedar ash catalyst is CaO. Under the optimal preparation conditions (first calcination temperature of 800°C, first calcination time 2 h, second calcination temperature 500°C, second calcination time 2 h), the prepared cedar ash catalyst can be used for transesterification and a better biodiesel yield of 91.52% can be achieved under catalyst concentration 7.0 wt%, reaction time 5 h, methanol/oil molar ratio 14:1 and reaction temperature 65°C. Cedar ash catalyst has good reusability performance. After four-time regeneration, the biodiesel yield can still reach 82.03%.

BDS and DS were prepared by burning biodiesel and $0^{\#}$ diesel, respectively, at room temperature and atmospheric pressure. BDS and DS are both chain aggregates, which are composed of a large number of spherical primary particles. The average diameter of the primary particles of the BDS was 35 nm, while that of the primary particles of the DS was 39 nm. BDS and DS both contained amorphous carbon and graphite. The degree of disorder of the DS ($I_D/I_G = 3.162$) exceeded that of the BDS ($I_D/I_G = 2.937$). Compared with the DS, the BDS had a higher C content and a lower content in O, H, N, and S. Both types of soot contained C—C, C—O—C, and C—OH functional groups, but BDS did not contain C=O functional groups.

BDS and DS significantly affected the viscosity of the LP, which increased exponentially with an increase of soot content. However, the influence of the BDS on the viscosity of the LP was greater than that of the DS, which is due to the fact that BDS more easily aggregated in the LP. The agglomeration mechanism of the BDS relative to that of the DS is that the surface energy of the BDS was higher and its lipophilic property was weaker, so that the agglomeration tendency of BDS in the LP was higher than that of the DS.

References

[1] Mardhiah HH, Ong HC, Masjuki HH, Lim S, Lee HV. A review on latest developments and future prospects of heterogeneous catalyst in biodiesel production from non-edible oils. Renew Sustain Energy Rev 2017;67:1225–36.

[2] Liu WW, Su YY, Zhang WD, Liu SQ, Xia CF. A study about analysis of glycerin content in biodiesel. Renew Energy 2005;121(3):14–20.

[3] Niu SL, Li H, Lu CM, Liu MQ, Huo MJ. Catalytic performance of papermaking white clay in the transesterification of peanut oil with methanol. J Fuel Chem Technol 2013;41(7):856–61.
[4] Sharma M, Khan AA, Puri SK, Tuli DK. Wood ash as a potential heterogeneous catalyst for biodiesel synthesis. Biomass Bioenergy 2012;41:94–106.
[5] Correia LM, Saboya RMA, Campelo NDS, Cecilia JA, Rodríguez-Castellón E, Cavalcante Jr CL, Vieira RS. Characterization of calcium oxide catalysts from natural sources and their application in the transesterification of sunflower oil. Bioresour Technol 2014;151:207–13.
[6] Tang ZX, Claveau D, Corcuff R, Belkacemi K, Arul J. Preparation of nano-CaO using thermal-decomposition method. Mater Lett 2008;62(14):2096–8.
[7] Tang Y, Wang SS, Liu Y, Zhang S, Gao XQ. Catalytic performance of modified CaO for biodiesel preparation. Acta Petrol Sin: Pet Process Section 2014;30(5):810–6.
[8] Yung BO, Gon S, Duck RC. Transesterification of tributyrin with methanol over calcium oxide catalysts prepared from various precursors. Fuel Process Technol 2009; 90(10):1252–8.
[9] Wang JX, Chen KT, Chen CC. Biodiesel production from soybean oil catalyzed by K_2SiO_3/C. Chinese J Catal 2011;32(10):1592–6.
[10] Jiang LH, Yan SL, Liang B. Heterogeneous solid base catalysts for biodiesel production. Ind Catal 2006;14(5):34–8.
[11] Guo XF, Chen J, Liu C. Production of the novel biodiesel over potassium floride and calcium oxide loaded kaolinite solid base catalyst. Chin J Appl Chem 2015;32(7):788–93.
[12] Li FJ, Li HQ, Wang LG, Cao Y. Waste carbide slag as a solid base catalyst for effective synthesis of biodiesel via transesterification of soybean oil with methanol. Fuel Process Technol 2015;131:421–9.
[13] Kotwal MS, Niphadkar PS, Deshpande SS, Bokade VV, Joshi P. Transesterification of sunflower oil catalyzed by fly ash-based solid catalysts. Fuel 2009;88(9):1773–8.
[14] Birla A, Singh B, Upadhyay SN, Sharma YC. Kinetics studies of synthesis of biodiesel from waste frying oil using heterogeneous catalysts derived from snail shell. Bioresour Technol 2012;106:95–100.
[15] Jaiyen S, Naree T, Ngamcharussrivichai C. Comparative study of natural dolomitic rock and waste mixed seashells as heterogeneous catalysts for the methanolysis of palm oil to biodiesel. Renew Energy 2015;74:433–40.
[16] Maneerung T, Kawi S, Wang CH. Biomass gasification bottom ash as a source of CaO catalyst for biodiesel production via transesterification of palm oil. Energ Conver Manage 2015;92:234–43.
[17] El Boulifi N, Bouaid A, Martinez M, Aracil J. Optimization and oxidative stability of biodiesel production from rice bran oil. Renew Energy 2013;53:141–7.
[18] Martino DS, Riccardo T, Lu PM, Santacesaria E. Heterogeneous catalysts for biodiesel production. Energy Fuel 2008;22:207–17.
[19] Rudnick LR. Lubricant additives: chemistry and applications. CRC Press, 2017.
[20] Salamanca M, Mondragón F, Agudelo JR, Benjumea P, Santamaría A. Variations in the chemical composition and morphology of soot induced by the unsaturation degree of biodiesel and a biodiesel blend. Combust Flame 2012;159(3):1100–8.
[21] Penchaliah R, Harvey TJ, Wood RJK, Nelson K, Powrie HEG. The effects of diesel contaminants on tribological performance on sliding steel on steel contacts. Proc Inst Mech Eng Part J J Eng Tribol 2011;225(8):779–97.
[22] Zhang B, Hu EZ, Liu TX, Hu XG. Characterization of morphology, structure and composition of soot particles from biomass fuel. CIESC J 2015;66(1):441–8.
[23] Hu EZ, Hu XG, Liu TX, Liu YM, Song RH, Chen YZ. Investigation of morphology, structure and composition of biomass-oil soot particles. Appl Surf Sci 2013;271(1):596–603.

[24] Clague ADH, Donnet JB, Wang TK, Peng JCM. A comparison of diesel engine soot with carbon black. Carbon 1999;37(10):1553–65.
[25] Wang Z, Li L, Wang JX, Reitz RD. Effect of biodiesel saturation on soot formation in diesel engines. Fuel 2016;175:240–8.
[26] Savic N, Rahman MM, Miljevic B, Saathoff H, Naumann KH, Leisner T, Riches J, Gupta B, Mota N, Ristovski ZD. Influence of biodiesel fuel composition on the morphology and microstructure of particles emitted from diesel engines. Carbon 2016;104:179–89.
[27] Sharma V, Uy D, Gangopadhyay A, Paxton WA, Sammut A, Ford MA, Aswath PBA. Structure and chemistry of crankcase and exhaust soot extracted from diesel engines. Carbon 2016;103:327–38.
[28] Zeidan M, Jia X, Williams RA, Wedlock DJ. Simulation of aggregation with applications of soot-laden lubricating fluids. Part Part Syst Char 2004;21(6):473–82.
[29] Mei DQ, Wang Z, Yuan YN, Ruan ZQ. Analysis of the features of the particulate matters generated from diesel engine fueled with biodiesel. Trans CSAE 2006;22(12):113–6.
[30] Li C, Hu XG. Green biomass ash catalyst for biodiesel synthesis. Acta Petrol Sin: Pet Process Section 2018;34(1):78–85.
[31] Kohse-Höinghaus K, Oßwald P, Cool TA, Kasper T, Hansen N, Qi F, Westbrook CK, Westmoreland PR. Biofuel combustion chemistry: from ethanol to biodiesel. Angew Chem Int Ed 2010;49(21):3572–97.
[32] Rohani B, Bae C. Morphology and nano-structure of soot in diesel spray and in engine exhaust. Fuel 2017;203:47–56.
[33] Merchan-Merchan W, Sanmiguel SG, McCollam S. Analysis of soot particles derived from biodiesels and diesel fuel air-flames. Fuel 2012;102:525–35.
[34] Joly-Pottuz L, Vacher B, Ohmae N, et al. Anti-wear and friction reducing mechanisms of carbon nano-onions as lubricant additives. Tribol Lett 2008;30(1):69–80.
[35] Schönfelder R, Avilés F, Bachmatiuk A, Cauich-Rodriguez JV, Knupfer M, Büchner B, Rümmeli MH. On the merits of Raman spectroscopy and thermogravimetric analysis to asses carbon nanotube structural modifications. Appl Phys A: Mater Sci Process 2012;106(4):843–52.
[36] Lapuerta M, Oliva F, Agudelo JR, Boehman AL. Effect of fuel on the soot nanostructure and consequences on loading and regeneration of diesel particulate filters. Combust Flame 2012;159(2):844–53.
[37] Salamanca M, Mondragón F, Agudelo JR, Santamaría A. Influence of palm oil biodiesel on the chemical and morphological characteristics of particulate matter emitted by a diesel engine. Atmos Environ 2012;62:220–7.
[38] Fan MM, Mao WT, Yang J, Zhang PB. Qualitative and quantitative analysis on fatty acid methyl ester (biodiesel). Chem Ind Eng Prog 2012;31(6):1373–8.
[39] Mirghani MES, Kabbashi NA, Alam MZ, Qudsieh IY, Ma'an FR. Rapid method for the determination of moisture content in biodiesel using FTIR spectroscopy. J Am Oil Chem Soc 2011;88(12):1897–904.
[40] Han C, Liu YC, Liu C, Ma JZ, He H. Influence of combustion conditions on hydrophilic properties and microstructure of flame soot. J Phys Chem A 2012;116(16):4129–36.
[41] Zotti M, Ferroni A, Calvini P. Microfungal biodeterioration of historic paper: preliminary FTIR and microbiological analyses. Int Biodeter Biodegr 2008;62(2):186–94.
[42] Omidvarbornaa H, Kumara A, Kim DS. Variation of diesel soot characteristics by different types and blends of biodiesel in a laboratory combustion chamber. Sci Total Environ 2016;544:450–9.
[43] Kirchner U, Scheer V, Vogt R. FTIR spectroscopic investigation of the mechanism and kinetics of the heterogeneous reactions of NO_2 and HNO_3 with soot. J Phys Chem A 2000;104(39):8908–15.

[44] Vander Wal RL, Bryg VM, Hays MD. XPS analysis of combustion aerosols for chemical composition, surface chemistry, and carbon chemical state. Anal Chem 2011;83(6):1924–30.
[45] Nieto-Gligorovski L, Net S, Gligorovski S, Zetzsch C, Jammoul A, D'Anna B, George C. Interactions of ozone with organic surface films in the presence of simulated sunlight: impact on wettability of aerosols. PCCP 2008;10:2964–71.
[46] Esangbedo C, Boehman AL, Perez JM. Characteristics of diesel engine soot that lead to excessive oil thickening. Tribol Int 2012;47:194–203.

CHAPTER 3

Tribological behavior of biodiesel soot

3.1 Effect of biodiesel soot on tribological behavior

3.1.1 Introduction

To mitigate global climate change, reduce dependence on fossil fuels, and increase national energy security, there is an urgent need to develop renewable fuels to replace fossil fuels [1]. Biodiesel is one of the promising alternative fuels for automotive engines. Produced from vegetable oils, animal fats, and waste cooking oil by transesterification, this renewable fuel is a mixture of fatty acid methyl esters [2]. Biodiesel's sulfur content is low and it does not contain aromatic hydrocarbons, which greatly decreases the content of sulfur compounds and carcinogens in exhaust gas, reducing environmental pollution. In addition, biodiesel is higher in oxygenated volume and burns more fully than petroleum diesel, resulting in higher combustion efficiency [3, 4]. As a result, many countries are taking measures to promote the use of biodiesel.

The problem of soot generation during the fuel combustion process is apparent [5]. Most of the soot generated during the combustion process is exhausted, but some can contaminate the lubricating oil within the sump as a result of blow-by gases [6]. This can be worsened when exhaust gas recirculation (EGR) technology is applied [7]. Soot particles are charged, hence they can easily agglomerate in lubricating oil. Soot particle aggregates in the lubricating oil can increase the viscosity of the oil and the particle size [8], which are both detrimental to the circulation of lubricating oil and can even lead to increased engine wear. Penchaliah et al. [9] analyzed the tribological behavior of soot in lubricating oils using a pin-on-disc tribometer under different conditions. It was found that soot could generate abrasion and starvation wear. Green et al. [10] reported that carbon black (acting as a soot simulant) leads to significant levels of abrasive wear when the content of carbon black in the lubricating oil is higher than 3%. Liu et al. [11] reported that biofuel soot could cause abrasive wear and corrosive wear. Previous research by this author found that the physicochemical and

Biodiesel Soot
https://doi.org/10.1016/B978-0-12-820553-2.00003-7

aggregation properties of biodiesel soot were significantly different from those of petrochemical diesel soot [12]. As a result, the tribological behaviors of biodiesel soot (BDS) in lubricating oils are different from those of diesel soot. However, few reports have described the effect of biodiesel soot on the tribological behavior of lubricating oil.

In this chapter, a four-ball tribometer was used to investigate the effect of BDS on the tribological performance of liquid paraffin (LP, acting as a lubricating base oil simulant). The tribological mechanism of BDS in LP was also studied. The aim was to lay a foundation for reducing the wear of BDS in lubricating oil and for providing an experimental basis for the extensive use of biodiesel in engines.

3.1.2 Experiments

3.1.2.1 Materials and samples preparation

Biodiesel was made from soybean oil and methanol by transesterification with self-made green biomass ash catalyst [13]. Table 3.1 shows the physical and chemical properties of biodiesel. BDS was collected from the combustion of the biodiesel using a self-made soot trap [12]. The collection of BDS comprised the following steps. Biodiesel was burned in a porcelain crucible at normal temperature and atmospheric pressure. A glass slide (soot capture slide) was held above the biodiesel flame until a soot layer a few micrometers thick was deposited [14]. This soot layer was then scraped off from the slide and was very carefully cleaned and collected in a glass bottle. Finally, the collected soot was dried under vacuum conditions at 120°C for 3 h and then ground to a fine granular powder. LP was purchased from Hengshui Diyi Petroleum Chemical Co. Ltd in China, with the physical and chemical properties shown in Table 3.1. The oil samples were LP + 1 wt% BDS, LP + 2 wt% BDS, LP + 3 wt% BDS, LP + 4 wt% BDS, and LP + 5 wt% BDS. The oil samples were stirred using a glass rod for 15 min. Finally, the oil samples were prepared by magnetic stirring for 1 h to reduce experimental deviation.

Table 3.1 Physical and chemical properties of biodiesel and liquid paraffin.

Items	Biodiesel	Liquid paraffin	Methods
Density at 20°C (kg/m^3)	875	882	GB/T 1884
Kinematic viscosity at 40°C (mm^2/s)	4.49	101	GB/T 265
Acid value (mg/g)	0.045	0.014	GB/T 264
Flash point (closed up) (°C)	168	210	GB/T 261
Water content (%)	Trace	Trace	GB/T 260

3.1.2.2 Tribological tests and characterization

Tribological tests were carried out on an MRS-10D four-ball tribometer produced by Jinan Shunmao Testing Machine Co. Ltd. in China. The balls (ϕ12.7 mm, HRC = 65 in hardness, and Ra = 0.012 μm) were made of GCr15 steel.

The primary particle morphology and internal structure of BDS were analyzed using a field-emission transmission electron microscope (FETEM, JEM-2100F). X-ray diffractometry (XRD, D/MAX2500V) was used to evaluate the BDS phase. The BDS chemical states were characterized via X-ray photoelectron spectroscopy (XPS; ESCALAB250). After friction, the rotational and stationary balls were washed with acetone. The morphologies of the wear zones and the elemental compositions on the surfaces were analyzed via 3D laser scanning microscopy (VK-X100, Keyence) and scanning electron microscopy coupled with energy dispersive spectroscopy (SEM/EDS, SU8010, Hitachi). The chemical states of the worn surfaces were investigated using XPS.

3.1.3 Characterization of biodiesel soot

Fig. 3.1 presents an FETEM image of BDS, showing the chain-like aggregates composed of spherical nanoparticles; the average diameter of a BDS particle was 35 nm by Nano Measurer analysis and calculation from Fig. 3.1A. Fig. 3.1B shows that the graphitic layers of the BDS were not continuous and were surrounded by amorphous phases [15]. These results were confirmed from the XRD spectra (Fig. 3.2) by the strong (002) and (101) diffraction peaks of Graphite-2H at 24.9 degrees and 43.5 degrees.

XPS spectra of BDS are shown in Fig. 3.3. As shown in Fig. 3.3A, the C_{1s} spectrum contained two main peaks. One is a strong peak at 284.8 eV, which corresponds to (C—C); the other is a small peak at 286.9, which indicates (C—O—C) and (C—O—H). As shown in Fig. 3.3B, the O_{1s} peak at 531.9 eV belongs to (C—O—C). The O_{1s} peak at 533.8 eV belongs to (C—O—H), suggesting that the oxygen-containing functional groups in BDS are (C—O—C) and (C—O—H), which made it possible for BDS to adsorb on friction surfaces. Interestingly, the BDS did not contain (C=O), possibly attributable to the (C=O) in the biodiesel being evolved as CO_2 or CO during combustion, because biodiesel contains more oxygen than does diesel [12, 16].

3.1.4 Effect of biodiesel soot on viscosity of liquid paraffin

Fig. 3.4 shows the variation of dynamic viscosity of LP contaminated with different contents of BDS at 20°C. The dynamic viscosity was measured on an NDJ-5S rotary viscometer, manufactured by Ningbo Weide

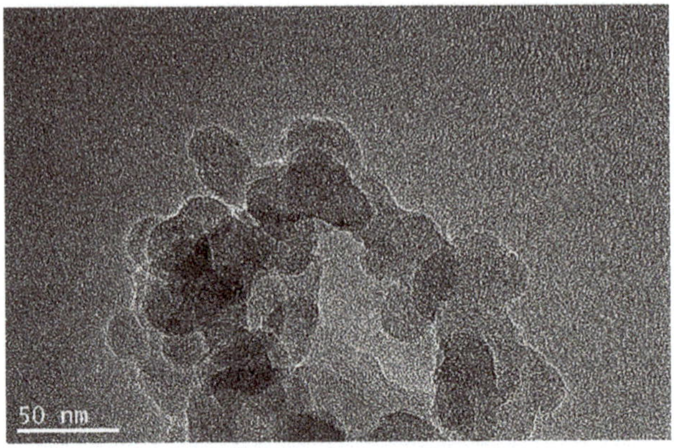

(A)

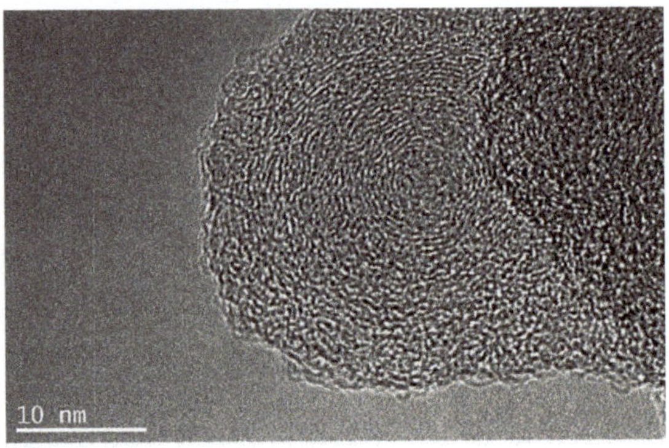

(B)

Fig. 3.1 FETEM images of BDS: (A) FETEM image of BDS (50 nm); (B) FETEM image of BDS (10 nm).

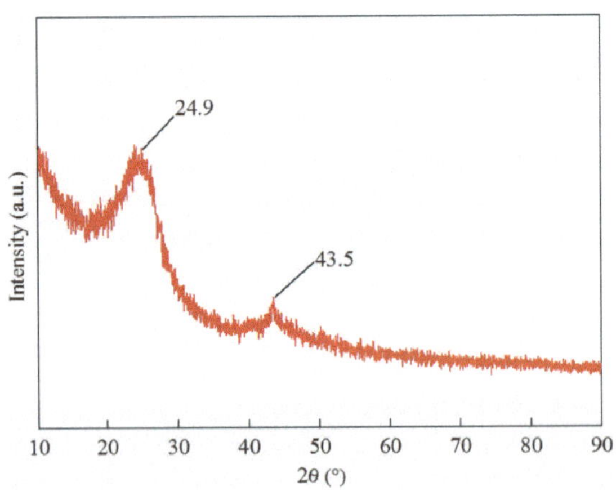

Fig. 3.2 XRD pattern of BDS.

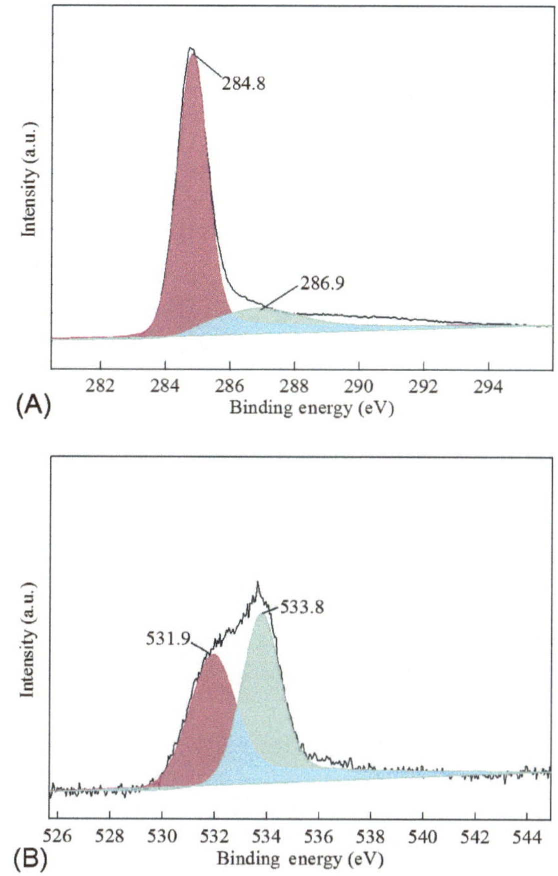

Fig. 3.3 XPS spectra of BDS: (A) C_{1s} spectra of BDS; (B) O_{1s} spectra of BDS.

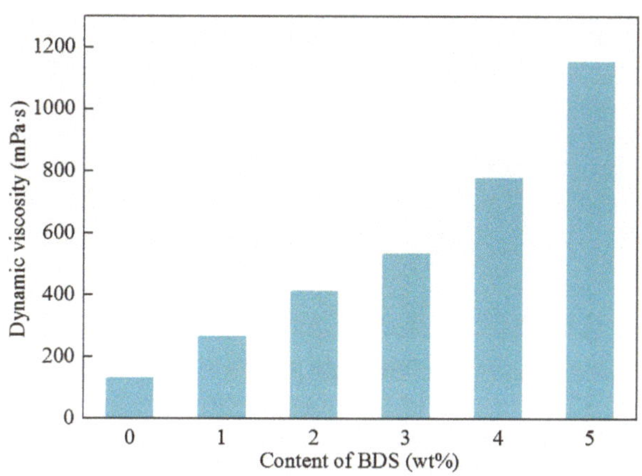

Fig. 3.4 Variation of dynamic viscosity with BDS content in contaminated LP.

Instrument Co. Ltd. in China. The viscosity change in LP caused by soot contamination can directly reflect the degree of soot aggregation [17]. The dynamic viscosity of LP showed significant increase with increasing BDS content. The results indicated that the content of BDS aggregates in LP increased with increasing BDS content [18].

3.1.5 Effect of biodiesel soot on the tribological behavior

Fig. 3.5 shows the effect of BDS content on the maximum nonseizure load (PB value) of LP. The PB value of LP presented an increasing trend with increasing content of BDS. When the content of BDS exceeded 3 wt%, the PB value of LP remained stable at 560 N. The increased PB value can be attributed to the dynamic viscosity of LP increasing with increasing content of BDS, which is beneficial for improvement of the bearing capacity of the oil film [19].

Fig. 3.6 shows the effects of BDS content on the average friction coefficient and average wear scar diameter (AWSD) of LP. The tribo-tests were performed on a four-ball tribometer at 196 N, 1450 rpm, room temperature (~20°C), and 30 min. As shown in Fig. 3.6A, with increasing content of BDS, the average friction coefficient of LP first decreased. However, when the content of BDS exceeded 1 wt%, the average friction coefficient of LP showed an increasing trend. This indicated that a small amount of BDS in LP is beneficial for improving the friction-reducing performance of LP. However, with increasing content of BDS, more BDS aggregates accumulate at the entraining edge of the tribocontact, resulting in partial starvation. Thus, a high content of BDS can degrade the friction-reduction properties

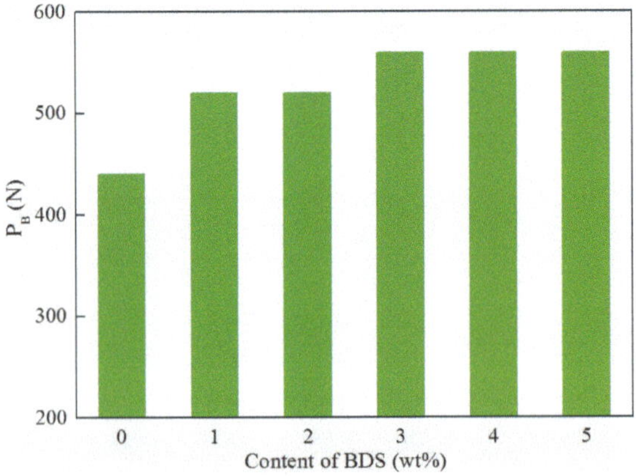

Fig. 3.5 Effect of BDS content on the maximum nonseizure load of LP.

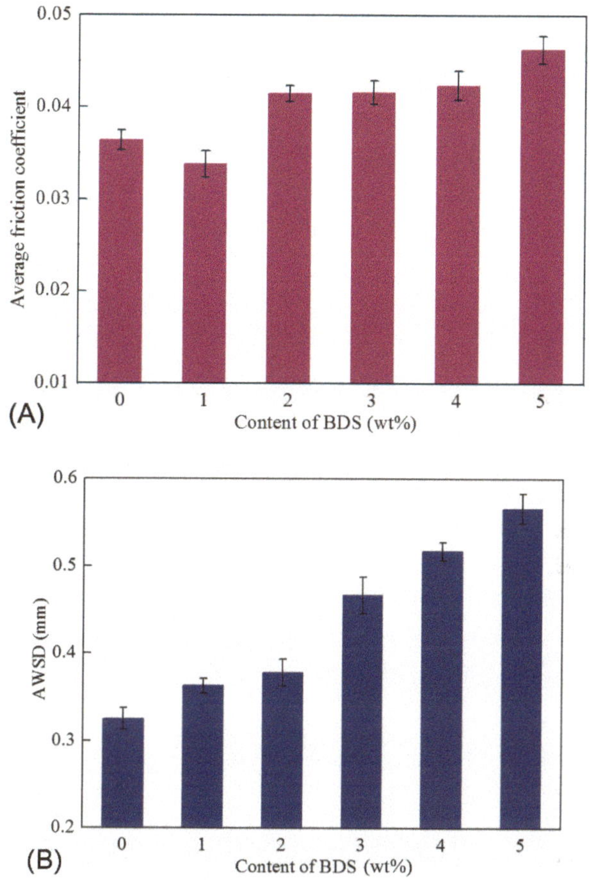

Fig. 3.6 Effect of BDS content on AWSD of LP: (A) Average friction coefficient; (B) AWSD.

of LP [20]. As shown in Fig. 3.6B, with increasing content of BDS, the AWSD of the LPincreased significantly. This indicates that the number of BDS aggregates in LP increased with increasing content of BDS, resulting in increased abrasive wear on the friction surfaces. Hence, BDS can reduce the wear resistance of LP.

3.2 Tribological mechanism of biodiesel soot

LP + 2 wt% BDS and LP + 5 wt% BDS represent levels of low and high soot in lubricating oil, respectively [21]. Hence, worn surfaces lubricated with LP, LP + 2 wt% BDS, and LP + 5 wt% BDS were selected to analyze the tribological mechanism. Fig. 3.7 shows the SEM image and the

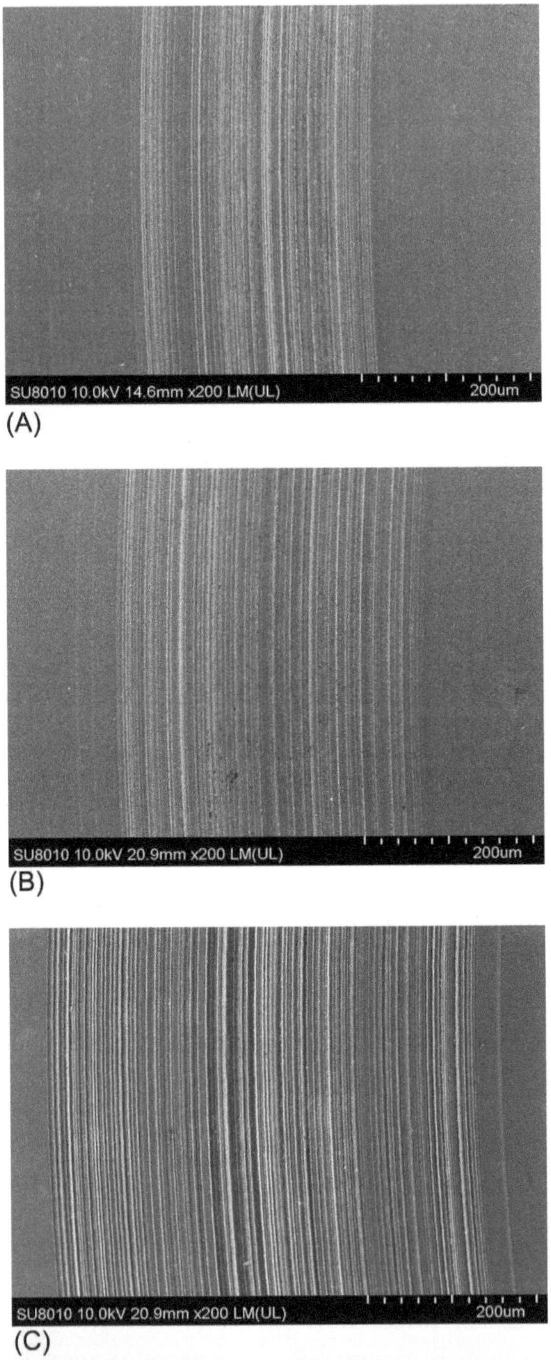

Fig. 3.7 SEM and the corresponding worn surface profiles of wear scars of rotational ball lubricated with different oil samples: (A) LP; (B) LP + 2% BDS; (C) LP + 5% BDS;

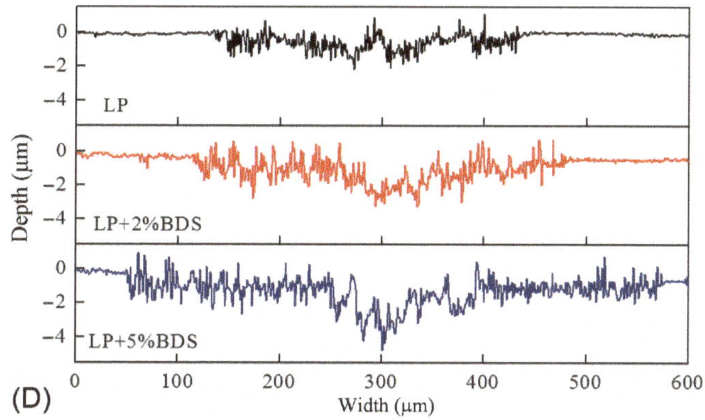

Fig. 3.7, cont'd (D) Worn surface profiles of wear scars of rotational ball.

corresponding worn surface profiles of wear scars of the rotational ball lubricated with different oil samples. Many furrows can be observed on the surface of the rotational ball after lubrication with LP in Fig. 3.7A. As shown in Fig. 3.7B and C, with increasing content of BDS, the rotational ball displays wider wear scars and more distinct furrows, which corresponds with the AWSD of the stationary ball in Fig. 3.6B. These results are confirmed by the worn surface profiles of wear scars of the rotational ball lubricated with different oil samples (shown in Fig. 3.7D). With LP, the furrows of the rotational ball are light and shallow. With increasing content of BDS, the furrows of the rotational ball become wider and deeper [22]. This supports the argument that the abrasive wear becomes more serious with increase in the content of BDS [23].

EDS spectra of the worn surfaces of the rotational balls lubricated with different oil samples are shown in Fig. 3.8. C, O, Fe, and Cr were detected in the worn surfaces of the rotational balls lubricated with different oil samples. The atomic content of C on the worn surfaces lubricated with LP, LP containing 2 wt% BDS, and LP containing 5 wt% BDS were 1.86%, 2.33%, and 3.20%, respectively. This means that C increased and O decreased on the worn surfaces with increasing content of BDS, indicating that more BDS aggregates were adsorbed on friction surfaces when the content of BDS increased, which is detrimental to the formation of lubricating oil films [11].

The C_{1s}, O_{1s}, and Fe_{2p} spectra of the worn surfaces of the rotational balls lubricated with different oil samples were recorded, as shown in Fig. 3.9. Fig. 3.9A–C shows the C_{1s} spectra of the worn surfaces. (C–H, C–C), C–O, and (C=O, O–C–O) peaks positioned at binding energies of 284.7 eV,

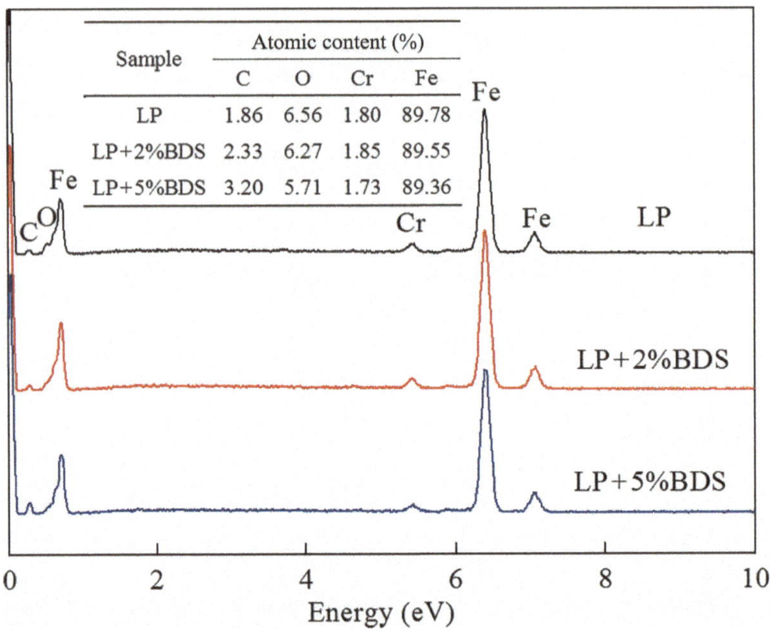

Fig. 3.8 EDS spectra of the worn surfaces of rotational balls lubricated with different oil samples.

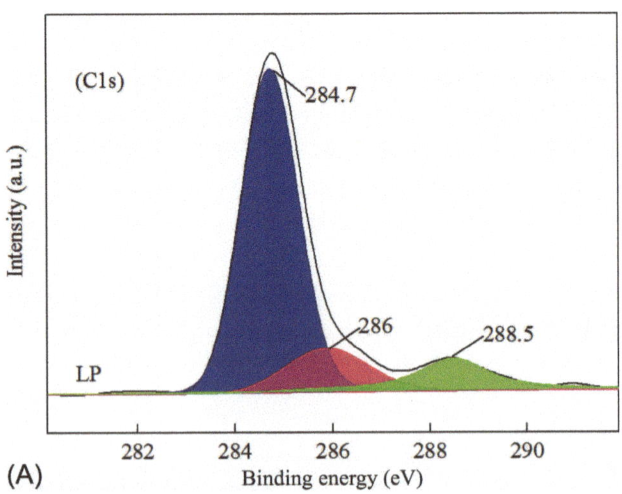

Fig. 3.9 C_{1s}, O_{1s} and Fe_{2p} spectra of the worn surfaces of rotational ball lubricated with different oil samples.

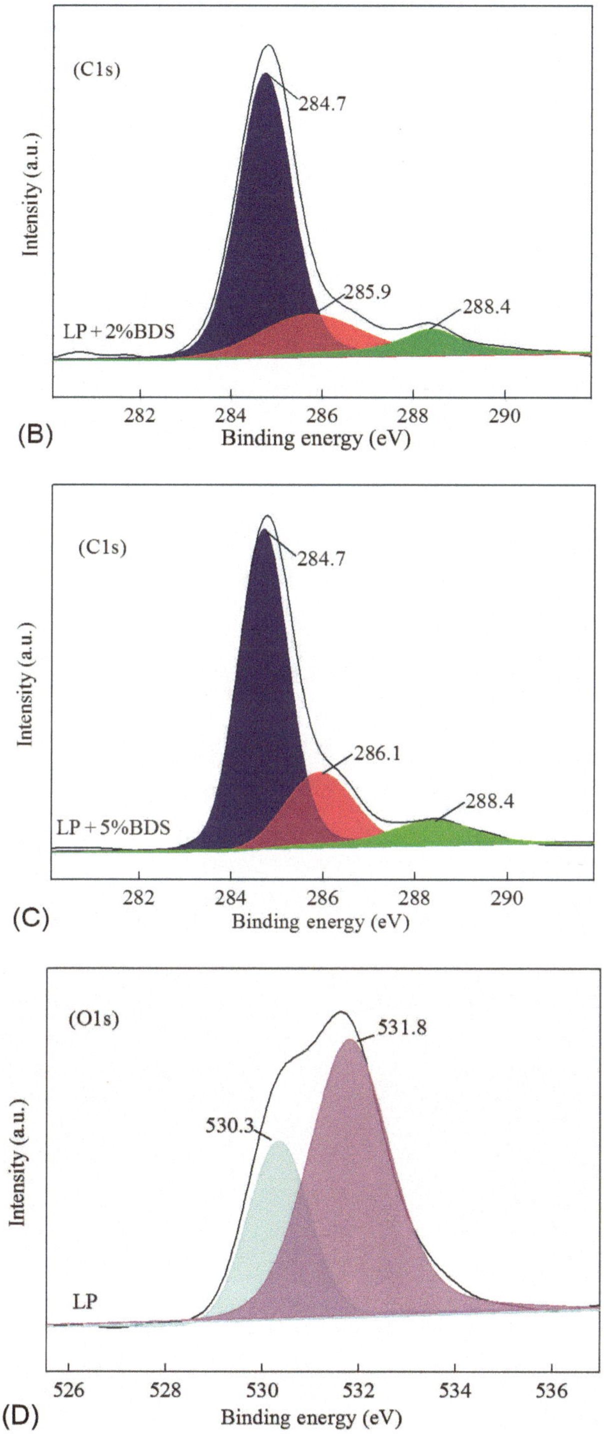

Fig. 3.9, cont'd

(Continued)

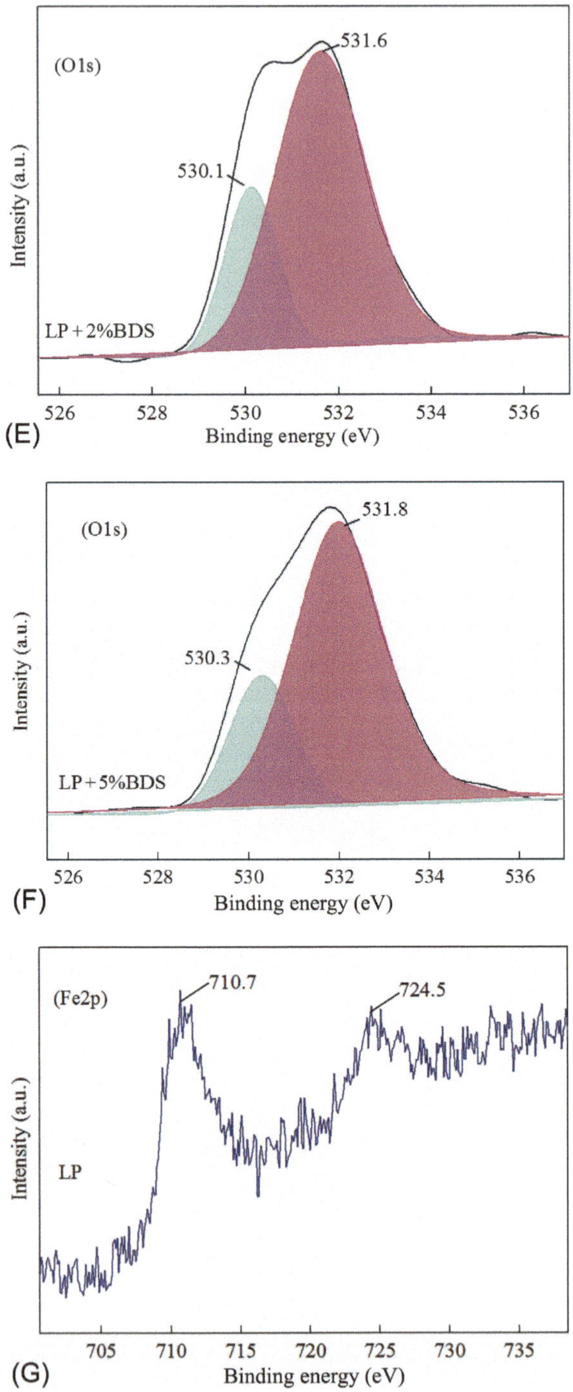
(O1s)
531.6
530.1
LP + 2%BDS
Intensity (a.u.)
526
528
530
532
534
536
Binding energy (eV)
(E)
(O1s)
531.8
530.3
LP + 5%BDS
Intensity (a.u.)
526
528
530
532
534
536
Binding energy (eV)
(F)
(Fe2p)
710.7
724.5
LP
Intensity (a.u.)
705
710
715
720
725
730
735
Binding energy (eV)
(G)

Fig. 3.9, cont'd

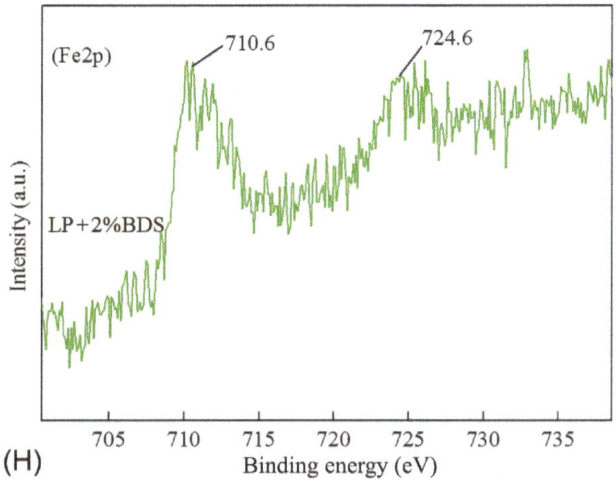

(H)

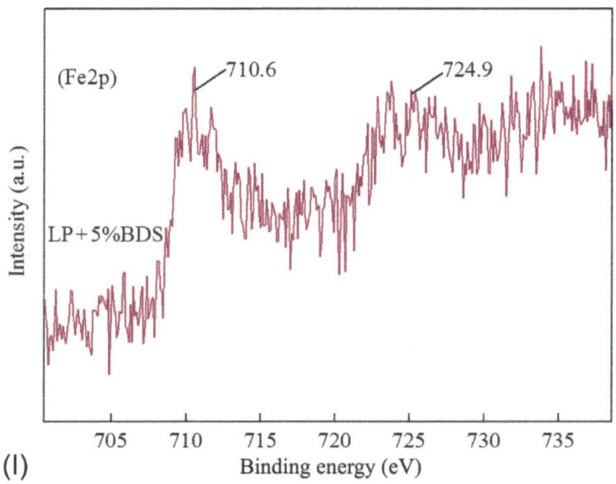

(I)

Fig. 3.9, cont'd

285.9–286.1 eV, and 288.4–288.5 eV, respectively [24], were observed in these three samples. Table 3.2 shows the functional group contents of carbon on the worn surfaces. The content of C—O increased obviously and the content of (C=O, O—C—O) decreased in the presence of BDS in LP, because the C—O from BDS adsorbed on the friction surfaces, suggesting that BDS can be adsorbed on rubbing surfaces. Fig. 3.9D–F shows the O_{1s} spectra of the worn surfaces. Metal oxide and (C—O, C=O, and —OH) peaks were positioned at binding energies of 530.1–530.3 eV and 531.6–531.8 eV, respectively. Table 3.3 shows functional group contents of oxygen on the worn surfaces. The content of (C—O, C=O and —OH) obviously

Table 3.2 Functional group contents of carbon on the worn surfaces.

Samples	Atom content (%)		
	(C—C, C—H)	C—O	(C=O, O—C—O)
LP	74	13.8	12.2
LP + 2% BDS	70.6	18.6	10.8
LP + 5% BDS	70.8	21.8	7.4

Table 3.3 Functional group contents of oxygen on the worn surfaces.

Samples	Atom content (%)	
	Metal oxide	(C—O, C=O and —OH)
LP	28.8	71.2
LP + 2% BDS	24.4	75.6
LP + 5% BDS	22.5	77.5

increased in the presence of BDS in LP, indicating that oxygen-containing functional groups of BDS adsorbed on friction surfaces [25]. Fig. 3.9G–I shows the Fe_{2p} spectra of the worn surfaces. $Fe_{2p3/2}$ and $Fe_{2p1/2}$ peaks positioned at binding energies of 710.6–710.7 eV and 724.5–724.9 eV, respectively, agree with Fe_2O_3. The results indicate that iron atoms were subject to oxidative reactions on the friction surfaces [26].

In general, the tribological mechanisms for BDS in LP can be divided into two types. When the BDS content in the LP was low, it could play an important role in reducing friction levels and act as a friction modifier [11,27]. At high content of BDS, however, more BDS aggregates were adsorbed at the interface between the rotational ball and stationary ball [28], impeding the effect of the lubricating oil film. In addition, abrasive wear became more serious as a result of the increase of BDS aggregates. This could lead to lowering the friction reduction and wear resistance of LP.

3.3 Conclusions

This research describes a study of the tribological behavior of BDS-contaminated LP. Chain-like aggregates of BDS were composed of spherical nanoparticles with an average diameter of 35 nm. The graphitic layers in BDS were not continuous and were surrounded by amorphous phases. The oxygen-containing functional groups in BDS were (C—O—C) and (C—O—H).

The dynamic viscosity of LP increased with increasing BDS content. BDS exerted a strong effect on the P_B value of LP. BDS decreased the wear resistance of LP. The friction-reduction property of LP improved when the content of BDS was below 1 wt%. However, when the BDS content exceeded 1 wt%, the BDS decreased the friction-reduction ability.

The tribological mechanisms can be summarized as follows: at a low BDS content, the BDS acts as a solid lubricant to lower the friction coefficient. However, at high BDS content, on one hand, more BDS aggregates are adsorbed on friction surfaces, thereby influencing the effect of the lubricating oil film. On the other hand, the high content increased the BDS aggregation and caused abrasive wear, which can lead to lowering the friction reduction and antiwear capability of LP.

References

[1] Veillette M, Giroir-Fendler A, Faucheux N, Heitz M. Esterification of free fatty acids with methanol to biodiesel using heterogeneous catalysts: from model acid oil to microalgae lipids. Chem Eng J 2017;308:101–9.

[2] Mahmudul HM, Hagos FY, Mamat R, Adam AA, Ishak WFW, Alenezi R. Production, characterization and performance of biodiesel as an alternative fuel in diesel engines—a review. Renew Sustain Energy Rev 2017;72:497–509.

[3] Özener O, Yüksek L, Ergenç AT, Özkan M. Effects of soybean biodiesel on a DI diesel engine performance, emission and combustion characteristics. Fuel 2014;115:875–83.

[4] Kohse-Höinghaus K, Oßwald P, Cool TA, Kasper T, Hansen N, Qi F, Westbrook CK, Westmoreland PR. Biofuel combustion chemistry: from ethanol to biodiesel. Angew Chem Int Ed 2010;49(21):3572–97.

[5] Cheng MT, Chen HJ, Young LH, Yang HH, Tsai YI, Wang LC, Lu JH, Chen CB. Carbonaceous composition changes of heavy-duty diesel engine particles in relation to biodiesels, after treatments and engine loads. J Hazard Mater 2015;297:234–40.

[6] Chiñas-castillo F, Spikes HA. The behavior of diluted sooted oils in lubricated contacts. Tribol Lett 2004;16(4):317–22.

[7] Tsolakis A, Megaritis A, Wyszynski ML, Theinnoi K. Engine performance and emissions of a diesel engine operating on diesel-RME (rapeseed methyl ester) blends with EGR (exhaust gas recirculation). Energy 2007;32(11):2072–80.

[8] Esangbedo C, Boehman AL, Perez JM. Characteristics of diesel engine soot that leads to excessive oil thickening. Tribol Int 2012;47:194–203.

[9] Penchaliah R, Harvey TJ, Wood RJK, Nelson K, Powrie HEG. The effects of diesel contaminants on tribological performance on sliding steel on steel contacts. Proc Inst Mech Eng Part J J Eng Tribol 2011;225(8):779–97.

[10] Green DA, Lewis R. Effect of soot on oil properties and wear of engine components. J Phys D Appl Phys 2007;40(18):5488–501.

[11] Liu TX, Hu EZ, Jin T, Xu YF, Yu HQ, Hu XG. Composition and structure as well as tribological performance of bio-fuel soot. Tribology 2014;34(4):379–86.

[12] Li C, Wu B, Liu TX, Liu YM, Hu XG. Physical-chemical and aggregation properties of biodiesel soot particles prepared at normal temperature and atmospheric pressure. Trans Chin Soc Agric Eng 2017;33(13):90–6.

[13] Li C, Hu XG. Green biomass ash catalyst for biodiesel synthesis. Acta Petrol Sin:Pet Process Section 2018;34(1):78–85.

[14] Deng X, Mammen L, Butt HJ, Vollmer D. Candle soot as a template for a transparent robust superamphiphobic coating. Science 2012;335(6064):67–70.
[15] Merchan-Merchan W, Sanmiguel SG, McCollam S. Analysis of soot particles derived from biodiesels and diesel fuel air-flames. Fuel 2012;102:525–35.
[16] Salamanca M, Mondragón F, Agudelo JR, Benjumea P, Santamaría A. Variations in the chemical composition and morphology of soot induced by the unsaturation degree of biodiesel and a biodiesel blend. Combust Flame 2012;159(3):1100–8.
[17] Zeidan M, Jia X, Williams RA, Wedlock DJ. Simulation of aggregation with applications of soot-laden lubricating fluids. Part Part Syst Char 2004;21(6):473–82.
[18] Liu TX, Song RH, Hu EZ, Xu YF, Hu XG. Effect of surface properties of fuel soot particles on viscosity of lubricating oil. J Chem Ind Eng 2015;66(10):4123–30.
[19] Hu EZ, Hu XG, Liu TX, Fang L, Dearn KD, Xu HM. The role of soot particles in the tribological behavior of engine lubricating oils. Wear 2013;304:152–61.
[20] Green DA, Lewis R, Dwyer-joyce RS. Wear effects and mechanisms of soot-contaminated automotive lubricants. Proc Inst Mech Eng Part J J Eng Tribol 2006;220(3):159–69.
[21] Liu TX, Hu EZ, Song RH, Zhang B, Hu XG. Dispersion characteristics of biofuel soot particles. J Chem Ind Eng 2015;66(4):1506–13.
[22] Su P, Xiong Y, Liu X, Yang H, Fan L. The influence of diesel soot on tribological behavior of cylinder liner/piston ring. Tribology 2017;37(2):206–11.
[23] Su P, Xiong Y, Liu X, Yang H. Morphology, structure and tribological behavior of diesel soot. Tribology 2017;37(1):83–9.
[24] Hu EZ, Dearn K, Yang BX, Song RH, Xu YF, Hu XG. Tribofilm formation and characterization of lubricating oils with biofuel soot and inorganic fluorides. Tribol Int 2017;107:163–72.
[25] Hu EZ. Triboligical behavior and friction-induced structure change mechanisms of biofuel soot particles [D]. Hefei: Hefei University of Technology; 2014.
[26] Xu YF, Hu XG, Yuan K, Zhu GL, Wang WZ. Friction and wear behaviors of catalytic methylesterified bio-oil. Tribol Int 2014;71:168–74.
[27] Liu CM, Nemoto S, Ogano S. Effect of soot properties in diesel engine oils on frictional characteristics. Tribol Trans 2003;46(1):12–8.
[28] Gautam M, Chitoor K, Durbha M, Summers JC. Effect of diesel soot contaminated oil on engine wear—investigation of novel oil formulations. Tribol Int 1999;432(12):687–99.

CHAPTER 4

Dispersion of biodiesel soot

4.1 Dispersion characteristics of biodiesel soot

4.1.1 Introduction

With the large consumption of fossil fuels worldwide, serious problems such as environmental degradation and energy shortages have arisen [1, 2]. Currently, there is an urgent need to develop environmentally friendly renewable fuels to replace fossil fuels. For automotive engines, biodiesel is currently one of the most promising alternative fuels [3]. Fatty acid glycerides from different sources, including animal fats and vegetable oils, can be prepared as biodiesel through transesterification. Compared with petrochemical diesel, biodiesel has the advantages of low carbon emissions and fewer harmful pollutants generated during combustion. Therefore, biodiesel has become an important research topic in the field of renewable energy [4, 5].

As an engine operates, if the air and fuel are not sufficiently mixed, the combustion proceeds unevenly and incompletely. Although the amount of soot produced by the combustion of biodiesel is smaller than that of petrochemical diesel, some soot particles still pass through the piston ring and enter the crankcase, thereby contaminating the lubricant [6–8]. With the widespread application of exhaust gas recirculation systems in combustion engines, the content of soot particles in lubricating oil is increasing [9]. Soot particles tend to agglomerate, which leads to an increase in the particle size of the agglomerated soot and in the viscosity of the contaminated oil [10]. These factors cause poor lubricant circulation and poor lubricant film formation, resulting in increased wear and even catastrophic failure [11]. Penchaliah et al. [12] studied the effects of diesel soot on the friction and wear behavior of steel–steel friction pairs and found that soot increased the viscosity of lubricants and caused abrasive wear. Green et al. [13, 14] found that as the soot content in lubricating oil increased, the wear volume on the friction interface increased. When the soot content in the lubricating oil was higher, the agglomerated soot particles caused obvious scratching on

Biodiesel Soot
https://doi.org/10.1016/B978-0-12-820553-2.00004-9

the surface of the friction pair. Therefore, the agglomerated soot contained in the lubricating oil of combustion engines is a serious problem.

To prevent soot agglomeration, chemical additives, such as dispersants and detergents, are usually added to lubricating oil to meet the performance requirements of combustion engines. Covitch et al. [15] compared different lubricant additive formulations on a Mack T-7 engine test and found that dispersants and high-alkalinity sulfonate detergents effectively inhibited lubricant thickening. Esangbedo et al. [16] found that the degree of graphitization and surface oxygen content in soot affected the soot-adsorption effect of dispersants. Liu et al. [17] found that succinimide dispersants significantly reduced the viscosity of lubricating oil polluted by biomass fuel soot and reduced the particle size of agglomerated biomass fuel soot. Previous research found that BDS and $0^{\#}$ diesel soot had significant differences in terms of their physical and chemical properties and agglomeration characteristics [10]. For the soot particles produced by the combustion of different fuels, specific dispersants and detergents must be selected to solve the soot-dispersion problems in lubricants. In the presence of dispersants and detergents, agglomerated BDS in lubricating oil may change its tribological characteristics due to changes in its particle size and other factors. There are few research reports on these aspects. Therefore, the work reported on in this chapter has chosen the most widely used polyisobutylene succinimide dispersant (PIBSI) and calcium salt detergents (alkyl calcium salicylate (ACS), high alkaline calcium sulfonate (HACS), and sulfuration calcium alkylphenol (SCA)), focusing on the effects of dispersants and detergents on the dispersion characteristics of BDS. We used a high-frequency reciprocating test rig to study the effects of the dispersion characteristics of BDS on the tribological behavior of LP (liquid paraffin, a simulant of lubricant base oil). This study provides a basis for the wide application of biodiesel in engines and the effect of BDS on engine wear.

4.1.2 Experiments

4.1.2.1 Materials and oil sample preparation

We used a specially made soot-capture device to obtain BDS from burned biodiesel [10]. The LP was obtained from Hengshui Diyi Petrochemical Co., Ltd.; the PIBSI, ACS, HACS, and SCA were obtained from Jinzhou Kangtai Lubricant Additive Co., Ltd.; and the acetone was obtained from Sinopec Group Chemical Reagent Co., Ltd.

We referred to the SH/T 0760-2005 Diesel Engine Oil Performance Evaluation Method (Mack T-8 method) to determine the mass fractions of the BDS in the LP, which were 2% and 5%, respectively; these were, respectively, used to represent low and high soot levels in the lubricating oil. According to the literature method of [18], the oil sample preparation process of the BDS + LP and the BDS + additives + LP was as follows: we added BDS with 2% or 5% mass fraction and dispersant/detergent with 0%–10% mass fraction to the LP to obtain oil samples (the BDS, dispersant, and detergent mass fractions were calculated based on oil sample benchmarks), after stirring with a glass rod. Finally, we used a laboratory high-shear emulsification machine (SG400, Shanghai Shanggui Fluid Equipment Co., Ltd.) in a 25°C water bath to cut for 30 min at 800 rev/min.

A rotational viscometer (NDJ-5S, Ningbo Weide Instrument Co., Ltd.) was used to investigate the effect of additives on the dynamic viscosity of the LP containing BDS at 25°C. The dynamic viscosity of the oil sample was measured to characterize the soot dispersion. The lower the dynamic viscosity, the better the dispersion.

4.1.2.2 Characterization

We used a zeta potentiometer (Nano-ZS90, Malvern) to measure the particle size distribution of the BDS in the LP. XPS (ESCALAB250, Thermo) was used to analyze the element content and element chemical states on the surface of the BDS.

4.1.3 Effect of additives on the dynamic viscosity of the LP with BDS

The effects of the additives (dispersants and detergents) on the viscosity of soot-contaminated lubricants can directly reflect the degree of soot dispersion in the lubricants and hence can be used to study soot-dispersion characteristics [16]. Fig. 4.1 shows the effect of the additive content on the dynamic viscosity of the BDS-containing LP. It can be seen from Fig. 4.1A that as the mass fraction of the PIBSI increased, the viscosity of the BDS-containing LP showed a significant decreasing trend. Among them, for the LP containing 2% BDS, the mass fraction of PIBSI was 1% and the viscosity of the dispersion system was reduced to 145 mPa·s. When the mass fraction of PIBSI was higher than 1%, the viscosity changed little. For the LP with 5% BDS, when the mass fraction of PIBSI was 2%, the viscosity of the dispersion system was reduced to 190 mPa·s. When the mass fraction of PIBSI was higher than 2%, the viscosity tended to be stable.

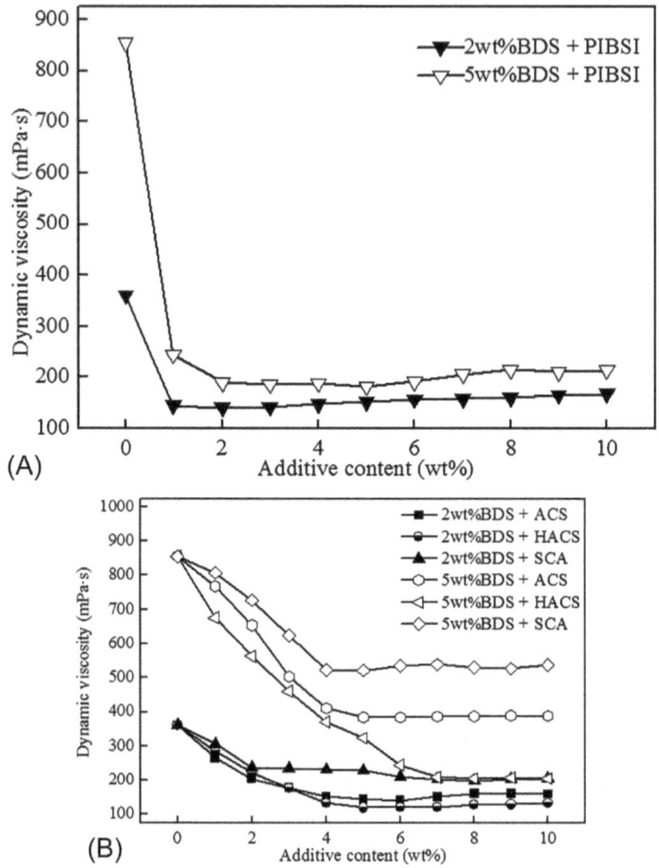

Fig. 4.1 Effect of additive content on the dynamic viscosity of LP with BDS: (A) PIBSI and (B) ACS, HACS and SCA.

In summary, when PIBSI was added to LP with different contents (2% and 5%) of BDS, the viscosity of the LP containing BDS was significantly reduced, indicating that PIBSI had a dispersal effect on LP with BDS.

From Fig. 4.1B, it can be seen that for the LP containing 2% BDS, as the SCA mass fraction increased, the viscosity of the oil sample gradually decreased. After a certain concentration, the viscosity was basically stable, but the viscosity of the oil sample decreased with an increase of the concentration of HACS and ACS, and the viscosity of the oil sample slightly increased after a certain concentration. When the concentrations of the three calcium salt detergents were less than 2%, the viscosity of the oil sample decreased similarly. At concentrations more than 2%, as the calcium salt detergent mass fraction increased, the viscosity of oil samples containing HACS decreased the most. When the HACS mass fraction was 5%, the viscosity reduced to a minimum, with a value of 120 mPa·s. For the LP containing 5% BDS, the viscosity of the oil sample decreased with an increase of the mass fraction of the calcium salt detergent, which became basically stable after exceeding a certain mass fraction. The dispersion effect of HACS was best for the same mass fraction of calcium salt detergent. When the HACS mass fraction was 7%, the viscosity was a minimum, with a value of 209 mPa·s. This indicates that the calcium salt detergent improved the dispersibility of the BDS in the LP. Among them, the dispersion effect of HACS on the LP with BDS was better than that of ACS and SCA. This is similar to the results of Huang [11], who found that sulfonate detergents are better than salicylate and salt of phenol in terms of solubilization and dispersion.

In engine lubricants, succinimide dispersants and high alkali value metal detergents are commonly used together and have synergistic effects [11]. Therefore, the influence of the combined use of PIBSI and HACS on the dispersion effect of BDS needs to be investigated. Fig. 4.2 shows the effect of the mass fractions of m_{HACS}/m_{PIBSI} and (PIBSI + HACS) on the dynamic viscosity of LP with BDS. It can be seen from Fig. 4.2A that the viscosity of the LP containing 2% or 5% BDS decreased first and then increased with an increase of m_{HACS}/m_{PIBSI}. When $m_{HACS}/m_{PIBSI} = 4$, the viscosity was a minimum. It can be seen from Fig. 4.2B that the viscosity of LP containing 2% BDS decreased significantly with an increase of the mass fraction of (PIBSI + HACS). When the mass fraction of (PIBSI + HACS) was 6%, the viscosity decreased to a minimum (101 mPa·s), while the LP with 5% BDS also had a similar trend. When the mass fraction of (PIBSI + HACS) was 9%, the viscosity decreased to a minimum (157 mPa·s). In addition, at the same mass fraction (PIBSI + HACS), the viscosity of LP with 2% BDS was lower than that with 5% BDS. The higher the BDS content, the higher the viscosity of the oil sample. It is shown in Fig. 4.1 that, compared to adding PIBSI or HACS separately, the viscosity reduction after adding (PIBSI + HACS) to LP with BDS was greater; that is, the dispersal effect of BDS in LP was better with the use of combined PIBSI and HACS.

4.1.4 Dispersion mechanism analysis

Using 5 mL samples of LP + 5% BDS + 2% PIBSI, LP + 5% BDS + 7% HACS, LP + 5% BDS + 9% (PIBSI + HACS) ($m_{HACS}/m_{PIBSI} = 4$, the same as in Section 4.1.3), and centrifugation and heptane washing methods [17], the BDS after additive adsorption was obtained and the C_{1s}, O_{1s}, N_{1s}, Ca_{2p}, and S_{2p} spectra before and after BDS adsorption of the additives

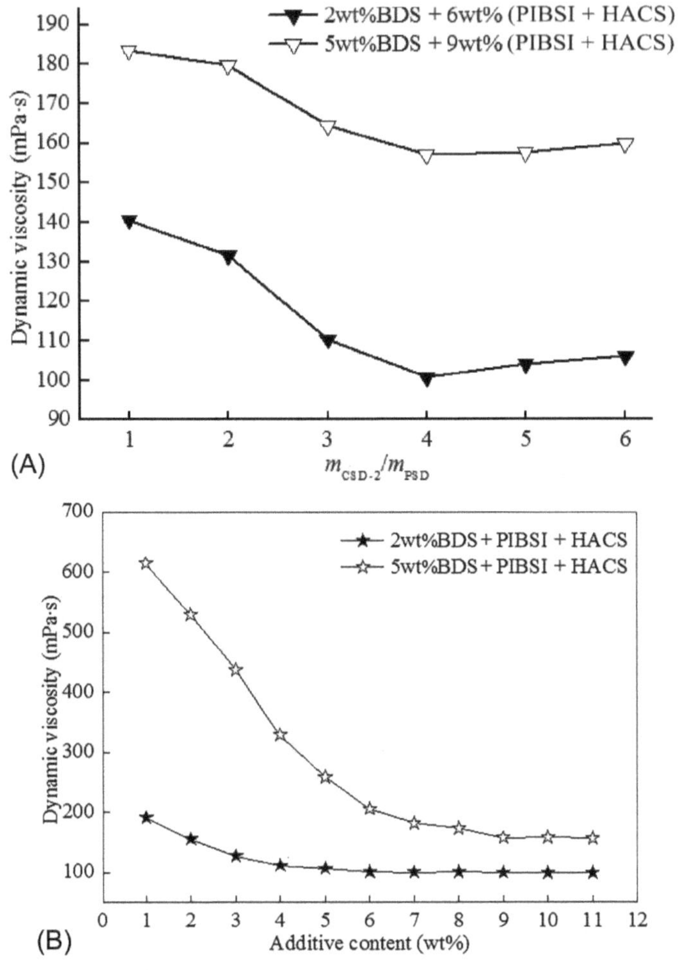

Fig. 4.2 Effect of m_{HACS}/m_{PIBSI} and (PIBSI + HACS) content on the dynamic viscosity of LP with BDS: (A) m_{HACS}/m_{PIBSI} and (B) (PIBSI + HACS) content.

were measured by XPS. As shown in Fig. 4.3, the peak in the C_{1s} spectrum at 284.8 eV is attributed to the C–C bond, and the 286.9 eV peak is the C–OH or C–O–C bond. The peak in the O_{1s} spectrum at 533.8 eV is due to the C–OH bond, and that around 531.9 eV is attributed to the C–O–C bond. These results indicate that the BDS before and after the addition of adsorption additives contained C–C, C–OH, and C–O–C bonds. After BDS interacted with PIBSI, N–C bonds (400.7 eV) appeared on the BDS surface. After BDS interacted with HACS, O–Ca–O bonds (348.5 and 352.3 eV) and S–O bonds (168.9 eV) appeared on the BDS surface [19].

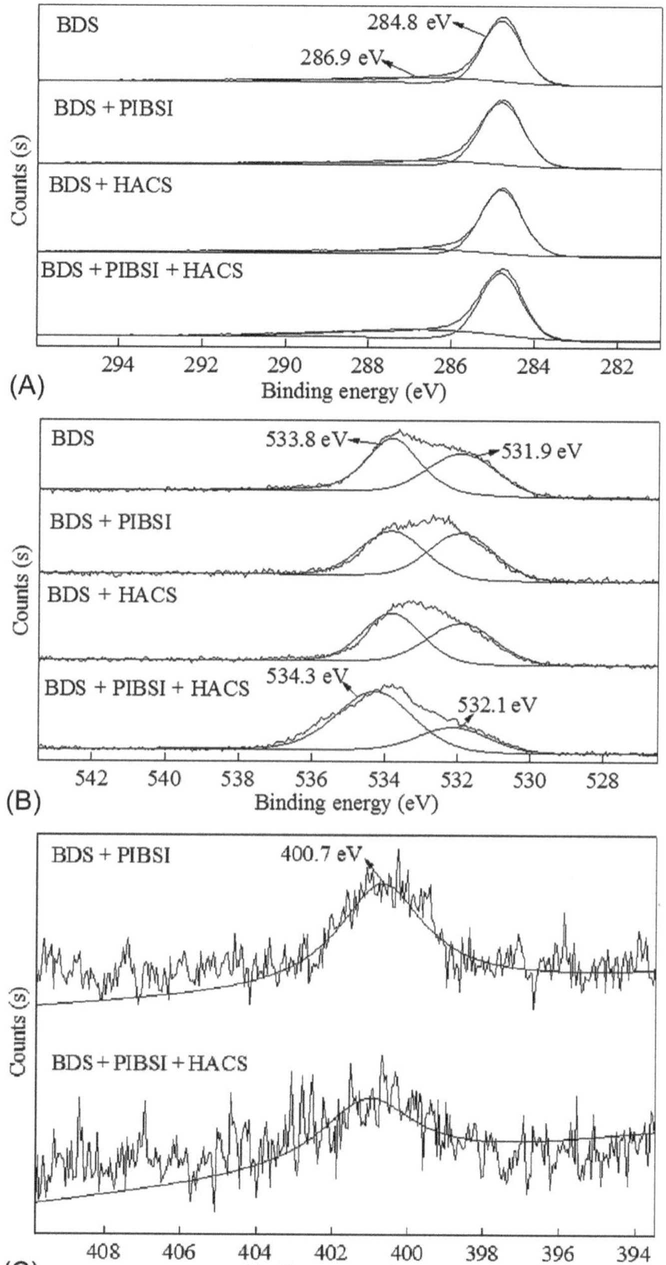

Fig. 4.3 See figure legend on next page

(Continued)

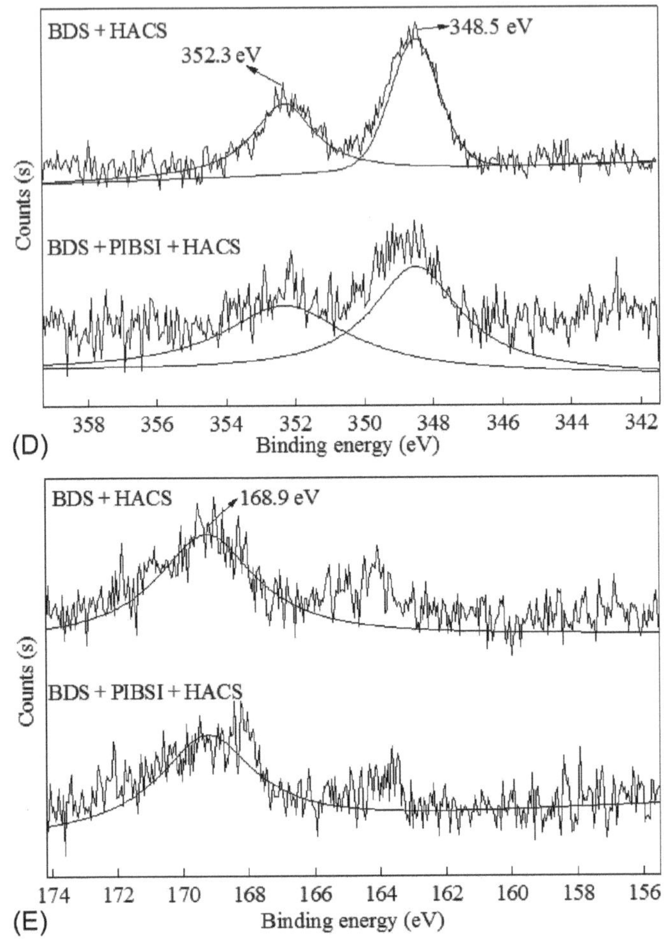

Fig. 4.3 C_{1s}, O_{1s}, N_{1s}, Ca_{2p}, and S_{2p} spectra of BDS adsorption additives before and after: (A) C_{1s}; (B) O_{1s}; (C) N_{1s}; (D) Ca_{2p}; (E) S_{2p}.

This indicates that the polar end of PIBSI or HACS had an adsorption effect with the oxygen-containing polar groups present on the surface of the BDS. After BDS interacted with (PIBSI + HACS), BDS showed the presence of N–C, O–Ca–O, and S–O bonds. This indicates that for the oil samples containing (PIBSI + HACS), both PIBSI and HACS adsorbed on the surface of the BDS.

As shown in Table 4.1, compared with the BDS, after BDS adsorbed PIBSI, the content in C and O decreased and N appeared. This shows that PIBSI adsorbed onto the surface of the BDS. After BDS adsorbed HACS or (PIBSI + HACS), in addition to the characteristic elements of the additive,

Table 4.1 XPS element content analysis of BDS adsorption additives before and after.

	Atomic content (%)				
Soot sample	**C**	**O**	**N**	**Ca**	**S**
BDS	92.61	7.39	–	–	–
BDS + PIBSI	90.99	7.03	1.98	–	–
BDS + HACS	90.1	8.55	–	0.91	0.44
BDS + PIBSI + HACS	86.67	10.36	1.74	0.85	0.38

the content in C decreased and the content in O increased, which may be related to the high total alkalinity value of HACS [19]. In addition, with reference to Fig. 4.3, it can be determined from calculations that the -OH content in BDS, BDS + PIBSI, BDS + HACS, and BDS + PIBSI + HACS were 3.86%, 3.43%, 4.39%, and 7.39%, respectively; that is, under the combined action of PIBSI and HACS, the large content in −OH indicates that the complex effects of PIBSI and HACS were favorable for the adsorption of the polar end of (PIBSI + HACS) and the hydroxyl and other oxygen-containing functional groups on the surface of the BDS through hydrogen bonding [20]. This is similar to the results of Ma et al. [21], who found that succinimide dispersants and calcium salt detergents formed complexing micelles after being compounded, which significantly enhanced the solubilization effect while maintaining good stability. Thus PIBSI and HACS had a positive synergistic effect.

Suspensions containing additives and BDS were prepared by ultrasonic dispersion and gradient dilution methods [10], and the particle size distributions of the agglomerated BDS before and after the addition of additives were measured using a zeta potentiometer. As shown in Fig. 4.4, the particle size distribution of the agglomerated BDS in the LP ranged from 122 to 396 nm. When PIBSI was added, the range was 59–190 nm, while when HACS was added, the range was 51–220 nm. Finally, when (PIBSI + HACS) was added, the range was 51–164 nm. The average particle size of the agglomerated BDS in the LP was 207.9 nm, while particles in the LP after PIBSI, HACS, and (PIBSI + HACS) were added measured an average of 101.2, 99.7, and 85.1 nm, respectively. After adding PIBSI and HACS separately and simultaneously, the average particle sizes of the aggregated BDS in the LP were significantly reduced, and the largest decrease was achieved after adding (PIBSI + HACS). This also indirectly proves that PIBSI and HACS had the best dispersal effect on the aggregated BDS in the LP. This is similar

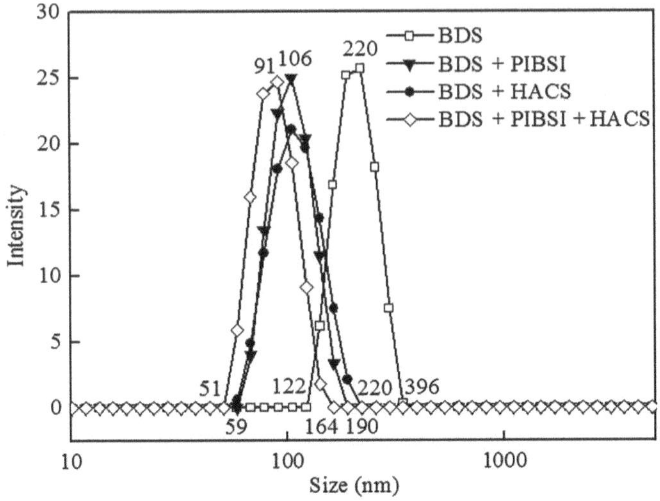

Fig. 4.4 Particle size distribution of BDS in LP dispersion.

to the results of the studies by Clague et al. [22] and Ratio et al. [23], who found that agglomerated soot easily agglomerates in oil to form a cluster structure. Dispersants promote a reduction of the binding force between the soot particles inside the cluster structure. Then the cluster structure dissociates and reduces the particle size of the agglomerated soot.

These conclusions indicate that in the BDS/PIBSI/LP system, PIBSI was associated with the surface of the BDS, and the lipophilicity of the hydrocarbon group (polyisobutylene) of the PIBSI caused the polyolefin chain to form a steric hindrance in the oil phase system, which ultimately suppressed the influence of the BDS [17]. In the BDS/HACS/LP system, HACS was associated with the surface of the BDS, thereby causing the BDS surface to have the charge; hence, the BDS particles repelled each other, thereby preventing BDS agglomeration [20]. In the BDS/(PIBSI + HACS)/LP system, (PIBSI + HACS) was associated with the surface of the BDS, which collectively prevented the BDS agglomerates from combining with each other. In addition, the combination of PIBSI and HACS had a solubilizing effect, which was beneficial for the polar end of (PIBSI + HACS) and oxygen-containing functional groups such as hydroxyl groups on the surface of BDS being adsorbed by hydrogen bonds, resulting in a smaller particle size of BDS in the LP. Thus, (PIBSI + HACS) had the effect of stabilizing the dispersion of BDS in LP.

4.2 Tribological behavior of biodiesel soot dispersion in lubricating oil

4.2.1 Experiments

4.2.1.1 Materials

A GCr15 steel ball and disk friction pair (steel ball, ϕ6 mm; hardness, 647–861 HV; surface roughness, 0.016 μm; and steel disk, ϕ 10 mm; thickness, 3 mm; hardness, 190–210 HV30; surface roughness 0.016 μm) were used for tribological tests.

Experimental method

The influence of the dispersion of the BDS on the tribological characteristics of the LP was studied with a high-frequency reciprocating tester (MGW-001, Jinan Yihua Tribology Testing Technology Co., Ltd.). The details are as follows: the oil sample was placed in the oil tank, and the contact interface between the steel ball and the steel disk was completely immersed in the oil sample. Reciprocating then occurred (frequency of 50 Hz, stroke length of 1 mm, 25°C, and test time of 75 min). At the end of the test, the coefficient of friction was recorded and the wear volume was measured [24].

We used scanning electron microscopy coupled with energy dispersive spectroscopy (SEM/EDS, JSM-6700F, Jeol) to analyze the morphology, element types, and content in the steel disk wear scar area. Also, 3D laser scanning microscopy (VK-X100, Keyence) was performed to visually analyze the microtopography and the wear surface profile of the steel disk wear scar area.

4.2.2 Effect of BDS dispersion on the tribological behavior of LP

The combined use of PIBSI and HACS significantly improved the dispersion of BDS in LP. Now we investigate the effect of the change in BDS dispersion caused by additives on the tribological properties of LP. Fig. 4.5 shows the effect of (PIBSI + HACS) content on the tribological properties of LP and LP with 2% BDS ($m_{HACS}/m_{PIBSI} = 4$). It is shown in Fig. 4.5A that the average friction coefficient of LP and LP with 2% BDS did not change much with an increase of the mass fraction of (PIBSI + HACS). When the mass fraction of (PIBSI + HACS) was the same, the average friction coefficient of LP with 2% BDS was lower than for just LP. This shows that (PIBSI + HACS) had little effect on the friction reduction performance of LP and LP with 2% BDS, and instead, it is BDS that had the effect of reducing the friction in LP [25]. It can be seen from Fig. 4.5B that

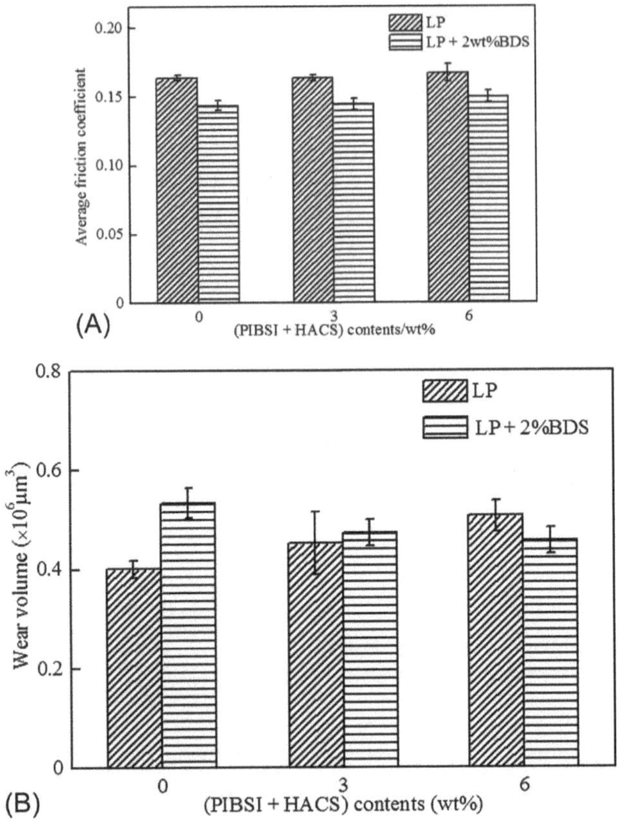

Fig. 4.5 Effect of (PIBSI + HACS) content on the tribological properties of LP and LP with 2% BDS ($m_{HACS}/m_{PIBSI} = 4$): (A) average friction coefficient and (B) wear volume.

the wear volume of the LP containing 2% BDS decreased with an increase of the mass fraction of (PIBSI + HACS), and the wear volume of the LP showed the opposite behavior. Under LP with 2% BDS, the wear volume was 5.33×10^5 μm^3. When different mass fractions (3% and 6%) (PIBSI + HACS) were added, the wear volume was reduced to 4.72×10^5 μm^3 (a decrease of 11.4%) and 4.56×10^5 μm^3 (a decrease of 14.4%). This shows that the addition of (PIBSI + HACS) improved the wear resistance of LP containing 2% BDS. This is similar to the results of studies by Mainwaring et al. [26] and Chiñas-Castillo et al. [27], who found that improving soot dispersion is beneficial to reducing wear.

Fig. 4.6 shows the effect of (PIBSI + HACS) content on the tribological properties of LP and LP with 5% BDS ($m_{HACS}/m_{PIBSI} = 4$). As shown in Fig. 4.6A, the influence of the mass fraction of (PIBSI + HACS) on the

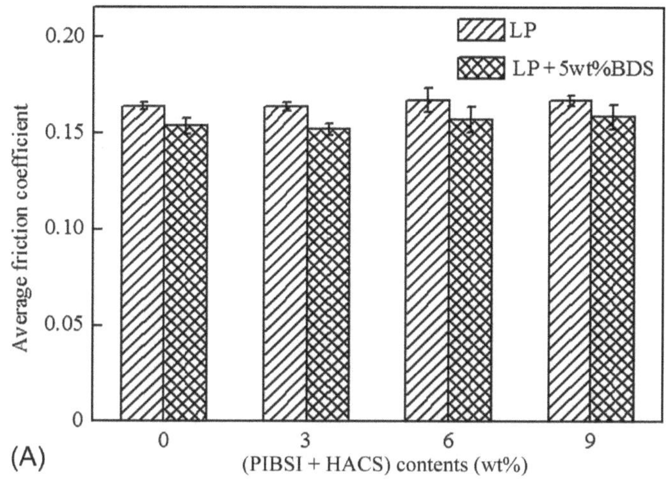

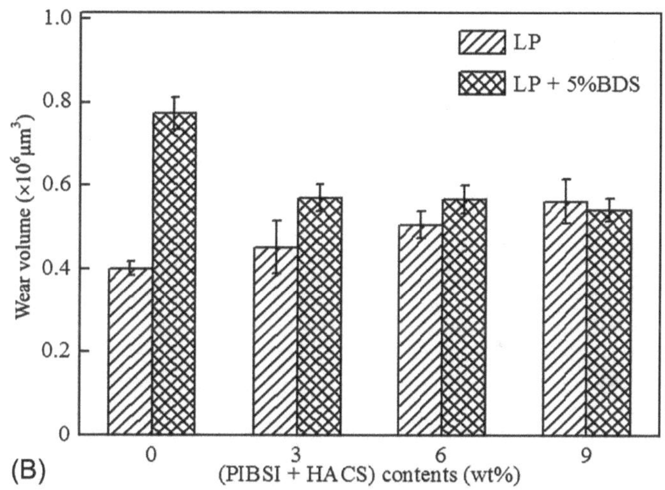

Fig. 4.6 Effect of (PIBSI + HACS) content on the tribological properties of LP and LP with 5% BDS ($m_{HACS}/m_{PIBSI} = 4$): (A) Average friction coefficient and (B) wear volume.

antifriction performance of LP containing 5% BDS is similar to that of LP containing 2% BDS. It is shown in Fig. 4.6B that the wear volume of the LP gradually increased as the mass fraction of (PIBSI + HACS) increased, while the wear volume of LP with 5% BDS decreased with an increase of the mass fraction of (PIBSI + HACS). When the (PIBSI + HACS) mass fraction increased from 3% to 9%, the reduction of wear volume gradually slowed down. With LP containing 5% BDS lubrication, the wear volume was 7.73×10^5 μm^3; when (PIBSI + HACS) with different mass fractions

(3%, 6%, and 9%) were added, the wear volume was $5.71 \times 10^5\ \mu m^3$ (a decrease of 26.1%), $5.68 \times 10^5\ \mu m^3$ (a decrease of 26.5%), and $5.42 \times 10^5\ \mu m^3$ (a decrease of 29.9%). With reference to Fig. 4.2B, it can be seen that as the (PIBSI + HACS) mass fraction increased from 0% to 9%, the viscosity of LP with 5% BDS gradually decreased and the dispersion became more uniform. Although the wear resistance caused by (PIBSI + HACS) was slightly reduced, in general, the wear resistance was gradually improved as the dispersion of the LP containing 5% BDS increased. This shows that (PIBSI + HACS) improved the antiwear performance of LP containing 5% BDS, and it had little effect on its antifriction performance.

4.2.3 Analysis of friction and wear

Fig. 4.7 shows SEM images (a′–c′), optical micrographs (a″–c″), and the corresponding worn surface profiles of wear scars on the steel disks lubricated with different oil samples. As seen in Fig. 4.7, in the SEM image, it can be observed that the surface of the steel disk was relatively flat when the LP lubrication was applied, although some shallow furrows appeared. There were a large number of furrows and pits on the surface of the steel disk when the LP containing 5% BDS was used. With the addition of 9% (PIBSI + HACS) to LP containing the 5% BDS, the number of furrows and pits on the surface of the steel disk was significantly reduced. According to the optical micrograph and the wear scar profile curve of the wear scar area of the steel disk, it can be found that with LP lubrication, the wear surface of the steel disk had narrower and shallower furrows and lighter wear. When using LP lubrication with 5% BDS, in the wear surface of the steel disk the furrows were

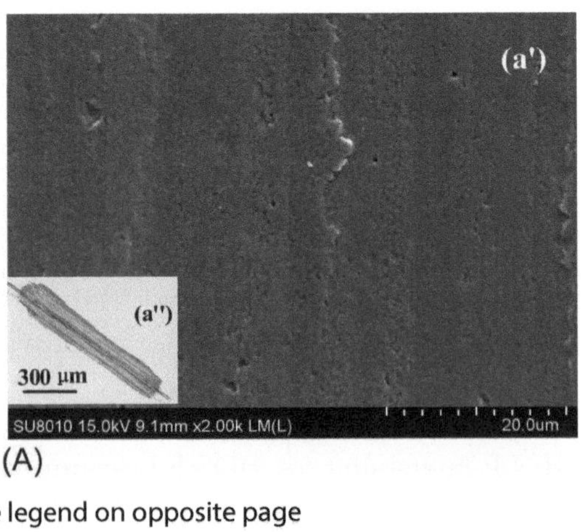

(A)

Fig. 4.7 See figure legend on opposite page

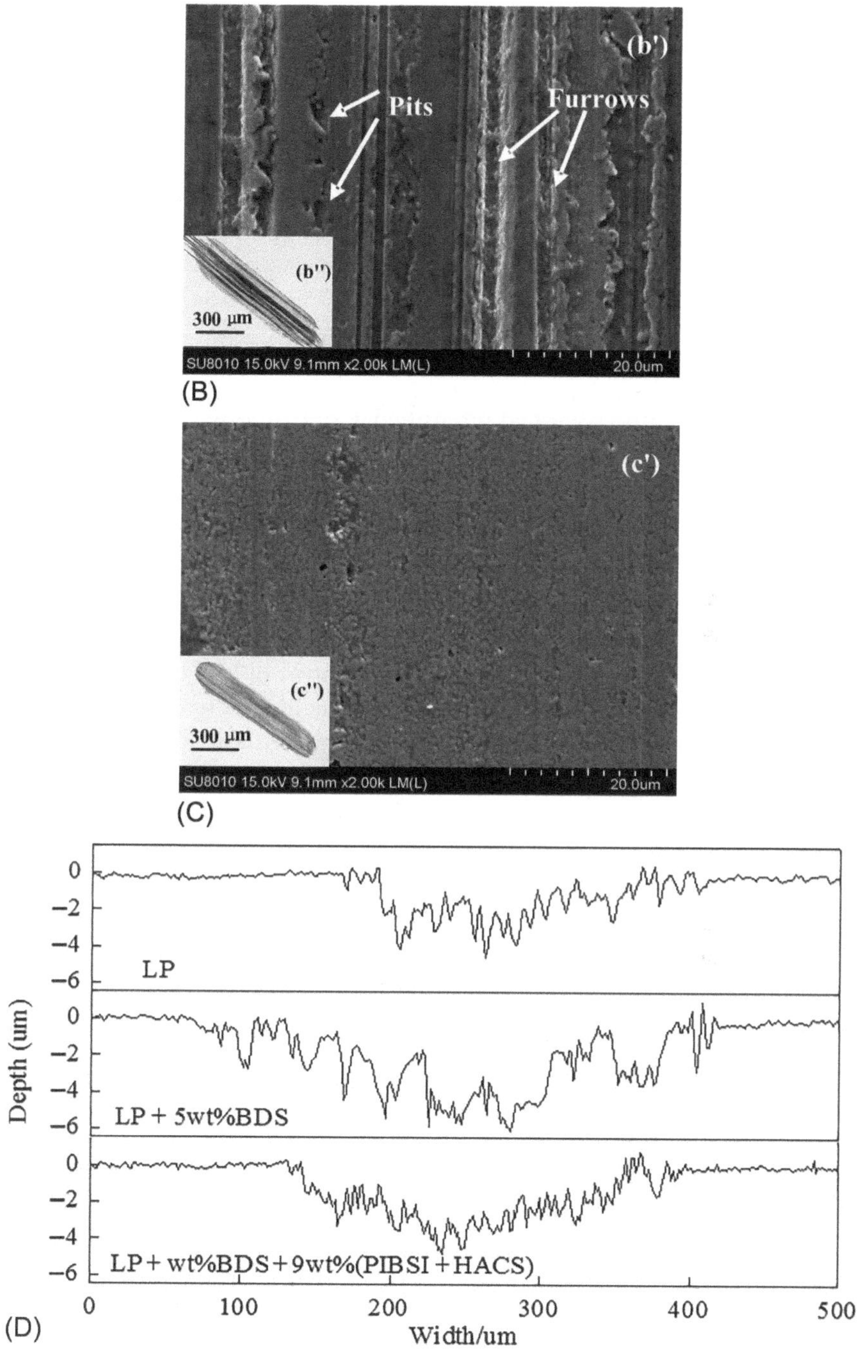

Fig. 4.7 SEM, optical micrographs, and the corresponding worn surface profiles of wear scars on the steel disks lubricated with different oil samples. (A) LP; (B) LP + 5% BDS; (C) LP + 5% BDS + 9% (PIBSI + HACS); (D) Worn surface profiles of wear scars on the steel disks.

Table 4.2 Energy spectrum analysis of wear scars on the steel disks lubricated with different oil samples.

Sample	Atomic content (%)						
	C	O	Cr	Fe	Ca	N	S
LP	2.81	10.49	3.09	83.61			
LP + 5% BDS	7.06	9.35	2.21	81.38	–	–	–
LP + 5% BDS + 9% (PIBSI + HACS)	2.68	15.40	2.00	76.53	0.84	2.24	0.31

wide and deep, and the wear had intensified. Compared with 5% BDS + LP lubrication, for LP containing 5% BDS and 9% (PIBSI + HACS), the wear scar surface of the steel disk became narrower and lighter, and the amount of wear reduced. These results correspond well with the SEM observations.

Further analysis of the wear scar region is shown in Table 4.2. The wear scar areas of the steel disks under the three oil samples all contained elements such as C, O, Fe, and Cr. Among them, Fe and Cr came from the matrix. With LP lubrication, in addition to C arising from the matrix, the other fraction of C arose from the carbon-containing organic compounds (hydrocarbons) which the LP adsorbed and deposited into a film during the friction process [28]. The O content mainly came from metal oxides (such as Fe_2O_3) that formed during the friction process. Compared with pure LP lubrication, the content in C was higher in the wear scar area of the steel disk when LP lubrication with 5% BDS was applied, and the content in O, Fe, and Cr was lower, which indicates that the agglomerated BDS adsorbed on the surface during the friction process [29]. The characteristic elements of the additive, such as Ca, N, S, and other elements, appeared on the steel disk wear scar areas when LP with 9% (PIBSI + HACS) and 5% BDS lubrication was used, which indicates that PIBSI and HACS participated in the formation of the lubricating film. It is worth mentioning that, compared with the LP lubrication containing BDS, the content of C in the wear scar area was reduced after adding (PIBSI + HACS), which proves that (PIBSI + HACS) dispersed, solubilized, and prevented the agglomeration of BDS, which was adsorbed on the surface, and subsequently reduced the abrasive wear of the BDS [21]. The increase in O content may have been due to the participation of HACS in the formation of the lubricating film.

Based on the results of these tests, it can be found that in the BDS/LP system, the particle sizes of the agglomerated BDS were large and could not be stably suspended in the LP. When the LP containing BDS was used as a lubricant, the agglomerated BDS easily deposited and adsorbed onto the surface during the friction process, which scratched the surface, causing abrasive wear [30]. In contrast, in the BDS/(PIBSI + HACS)/LP system, (PIBSI + HACS) was associated with the surface of the BDS, which inhibited mutual bonding between the agglomerated BDS and stopped the agglomerated BDS from getting bigger. At the same time, the solubilization of PIBSI and HACS kept the BDS suspended in the oil and inhibited the adsorption of BDS onto the surface [21]. Therefore, when the LP containing BDS and (PIBSI + HACS) was used as a lubricant, the particle sizes of the agglomerated BDS were small and stably dispersed in the LP, which reduced the amount of scratching on the surface by the agglomerated BDS and improved the wear resistance of LP containing BDS.

4.3 Conclusions

PIBSI can reduce the dynamic viscosity of the BDS-contaminated LP. Compared with three calcium salt detergents, HACS had a better dispersion effect on the BDS in the LP. The BDS surface had the highest −OH content because of the addition of (PIBSI + HACS), and the average particle size of the agglomerated BDS in the LP was the smallest. (PIBSI + HACS) had the best dispersion effect on BDS in the LP, and PIBSI and HACS combined had a synergistic effect.

When 6% (PIBSI + HACS) was added to LP containing 2% BDS, the wear volume reduced from 5.33×10^5 μm^3 to 4.56×10^5 μm^3, which was a decrease of 14.4%. When 9% (PIBSI + HACS) was added to the LP containing 5% BDS, the wear volume was reduced from 7.73×10^5 μm^3 to 5.42×10^5 μm^3, i.e., a decrease of 29.9%, indicating that (PIBSI + HACS) can improve the wear resistance of LP containing BDS.

(PIBSI + HACS) adsorbed onto the surface of BDS through hydrogen bonding, which hindered BDS agglomeration in the LP. (PIBSI + HACS) dispersed BDS and inhibited the adsorption of BDS onto the surface, which improved the antiwear performance of LP containing BDS.

References

[1] Chen CY, Lee WJ, Wang LC, Chang YC, Yang HH, Young LH, Lu JH, Tsai YI, Cheng MT, Mwangi JK. Impact of high soot-loaded and regenerated diesel particulate filters on the emissions of persistent organic pollutants from a diesel engine fueled with waste cooking oil-based biodiesel. Appl Energy 2017;191:35–43.

[2] Deng XL, Ye SS, Cao ZK, Zhang NW, Zheng JB, Chen BH. Effect of Ag loading on soot oxidation for Ag/$Ce_{0.75}Zr_{0.25}O_2$ catalysts. CIESC J 2017;68(8):3064–70.

[3] Jung Y, Hwang J, Bae C. Assessment of particulate matter in exhaust gas for biodiesel and diesel under conventional and low temperature combustion in a compression ignition engine. Fuel 2016;165:413–24.
[4] Cheng TC, Simone H. Flame structure, spectroscopy and emissions quantification of rapeseed biodiesel under model gas turbine conditions. Appl Energy 2017;185:1383–92.
[5] Datta A, Mandal BK. Engine performance, combustion and emission characteristics of a compression ignition engine operating on different biodiesel-alcohol blends. Energy 2017;125:470–83.
[6] Zhang YL, Zhang RL, Rao LZ, Kim D, Kook S. The influence of a large methyl ester on in-flame soot particle structures in a small-bore diesel engine. Fuel 2017;194:423–35.
[7] Zhang B, Hu EZ, Liu TX, Hu XG. Characterization of morphology, structure and composition of soot particles from biomass fuel. CIESC J 2015;66(1):441–8.
[8] Hu EZ, Dearn K, Yang BX, Song RH, Xu YF, Hu XG. Tribofilm formation and characterization of lubricating oils with biofuel soot and inorganic fluorides. Tribol Int 2017;107:163–72.
[9] Tsolakis A, Megaritis A, Wyszynski ML, Theinnoi K. Engine performance and emissions of a diesel engine operating on diesel-RME (rapeseed methyl ester) blends with EGR (exhaust gas recirculation). Energy 2007;32(11):2072–80.
[10] Li C, Wu B, Liu TX, Liu YM, Hu XG. Physical-chemical and aggregation properties of biodiesel soot particles prepared at normal temperature and atmospheric pressure. Trans Chin Soc Agric Eng 2017;33(13):90–6.
[11] Huang WX. Lubricant additives properties and applications. Beijing: Chemical Petrochemical Press; 2012. p. 41–54.
[12] Penchaliah R, Harvey TJ, Wood RJK, Nelson K, Powrie HEG. The effects of diesel contaminants on tribological performance on sliding steel on steel contacts. Proc Inst Mech Eng Part J J Eng Tribol 2011;225(8):779–97.
[13] Green DA, Lewis R. Effect of soot on oil properties and wear of engine components. J Phys D Appl Phys 2007;40(18):5488–501.
[14] Green DA, Lewis R, Dwyer-Joyce RS. Wear effects and mechanisms of soot-contaminated automotive lubricants. Proc Inst Mech Eng Part J J Eng Tribol 2006;220(3):159–69.
[15] Covitch MJ, Humphrey BK, Ripple DE. Oil thickening in the Mack T-7 engine test-fuel effects and the influence of lubricant additives on soot aggregation. SAE Pap 1985;852126.
[16] Esangbedo C, Boehman AL, Perez JM. Characteristics of diesel engine soot that lead to excessive oil thickening. Tribol Int 2012;47:194–203.
[17] Liu TX, Hu EZ, Song RH, Zhang B, Hu XG. Dispersion characteristics of biofuel soot particles. CIESC J 2015;66(4):1506–13.
[18] Liu TX, Song RH, Hu EZ, Xu YF, Hu XG. Effect of surface properties of fuel soot particles on viscosity of lubricating oil. CIESC J 2015;66(10):4123–30.
[19] Liu TX, Song RH, Liu YM, Xu YF, Hu XG. Effect of calcium salt detergent on dispersity of biofuel soot. CIESC J 2016;67(7):3032–9.
[20] Rudnick LR. Lubricant additives chemistry and applications. Beijing: Chemical Petrochemical Press; 2006. p. 95–118.
[21] Ma LC. Interactions between the additives used in universal engine oil. J Xi'an Shiyou Univ (Nat Sci Ed) 2007;22(4):84–8.
[22] Clague ADH, Donnet JB, Wang TK, Peng JCM. A comparison of diesel engine soot with carbon black. Carbon 1999;37(10):1553–65.
[23] Ratio M, Castle RC, Bovington CH, Spikes HA. The Influence of soot and dispersant on ZDDP film thickness and friction. Lubr Sci 2004;17(1):25–43.
[24] Su P, Xiong Y, Liu X, Yang H. Morphology, structure and tribological behavior of diesel soot. Tribology 2017;37(1):83–9.
[25] Hu EZ, Hu XG, Liu TX, Song RH, Dearn KD, Xu HM. Effect of TiF_3 catalyst on the tribological properties of carbon black-contaminated engine oils. Wear 2013;305(1-2):166–76.

[26] Mainwaring R. Soot and wear in heavy duty diesel engines. SAE Pap 1997;971631.
[27] Chinas-Castillo F, Spikes HA. The behavior of diluted sooted oils in lubricated contacts. Tribol Lett 2004;16(4):317–22.
[28] Xu YF, Hu XG, Yu HQ, HU LR. Lubrication performance and mechanism of bio-oils made from typical microalgae. Tribology 2014;34(3):291–6.
[29] Su P, Xiong Y, Liu X, Yang H, Fan LJ. The Influence of diesel soot on tribological behavior of cylinder liner/piston ring. Tribology 2017;37(2):206–11.
[30] Salehi FM, Khaemba DN, Morina A, Neville A. Corrosive–abrasive wear induced by soot in boundary lubrication regime. Tribol Lett 2016;63(2):19.

CHAPTER 5

Active control of tribological behavior of biodiesel soot

5.1 Overview of tribological active control

5.1.1 Introduction

A great deal of research has been carried out on the topic of actively controlling tribological performance [1–3], mainly concerned with the implementation of solid-liquid composite lubrication and external field control lubrication technology. In a solid-liquid composite lubrication system, it is common to add various nanoparticles (such as nanocopper, nanomolybdenum disulfide, nanoalumina, nanodiamond, and titanium dioxide) to lubricating oil or grease, where they play a role in the friction interface, improving the antiwear capability and friction reduction of the lubricating oil [4–6]. However, some problems can arise with this technique, such as difficulties with the dispersion of the nanoparticles, reaching the friction interface of the nanoparticles, and stable control of the interaction between the nanoparticles and the friction interface.

These issues have led to the use of external fields to actively control the lubrication, friction, or wear behaviors. Technologies such as magnetic ferrofluid lubrication, electrorheological fluids lubrication, and liquid crystal lubrication have been used since the 1930s, and gradually active control of the lubrication process and state has been realized [7]. At the same time, researchers have also pursued the active control of the friction and wear processes of the friction pairs in the machine, so as to reduce the wear of mechanical parts, save energy and material consumption, and reduce the machine cost, which extends the service life of the machine. Controlling the friction and wear processes by electric fields originated in the 1950s, and the control technology of magnetic fields originated from the research of Indian scholars in 1968. Ultrasonic active control technology started with research on the ultrasonic motor and guide rail of Japanese scholars in the 1990s. Its origin can be traced back to the beginning stages of research on ultrasonic waves in physics. At that time, people began to experiment with and study the influence of ultrasonics on wear.

Biodiesel Soot
https://doi.org/10.1016/B978-0-12-820553-2.00005-0

5.1.2 Electric field control

As early as 1953, British scholars studied friction under the action of electric fields. The influence of an external electric field on friction coefficient and wear rate was first studied; on the one hand, the wear of an electric contact friction pair was examined in an engineering analysis, and, on the other hand, active electric field control of the friction and wear of other friction pairs was studied. Hurricks [8] studied the friction between aluminum and carbon black rubber, and the friction coefficient increased after applying a voltage. Yamamoto et al. [9] found that the surface current affected the formation of the surface film. The formation of a boundary film on the surface was beneficial in improving the antibonding ability of the friction pair. However, when the voltage was too high, the surface film formation was decreased and the antiadhesive ability was reduced. The friction coefficient increased when an external voltage was applied but decreased when a reverse voltage was applied. The running in of the sliding bearing was stable after applying a voltage, and the friction coefficient dropped rapidly during the running in. These results showed that the wear volume of the ball disk friction pair was small at low voltage and low current. Yamamoto et al. [10] also studied the relationship between tribological properties, self-generated voltage, and magnetic field using a pin-disk tribometer. It was believed that eliminating the current between relative sliding interfaces could reduce friction and wear, and the friction coefficient could be reduced by 35%.

Zhai et al. investigated the friction and wear properties of steel-steel friction pairs lubricated with different lubricants and the effect of an external electric field on the friction and wear properties [11]. The results showed that applying voltage, especially changing the polarity of the voltage, could greatly change the friction and wear properties of friction pairs, while the effect of an electric field and its polarity on the friction and wear properties depended on the properties of the lubricant and boundary lubrication film. The friction coefficient was the smallest when the applied voltage and the self-generated potential of friction were offset, which was 12%–26% lower than when there was no applied voltage. At this time, the applied voltage could prevent the flow of electrons at the interface and reduced friction. The friction force of the metal friction pair was affected by the self-generated potential, the solubility of the friction metal, and its crystallization. Under dry friction, the surface film of the metal had an obvious influence on its frictional behavior and surface self-generated potential, and the influence of metal abrasive on the frictional behavior and spontaneous electric potential depended on the adhesion characteristics of the abrasive particles and the friction pairs.

According to the tribological results of plate-to-plate friction pairs by Meng et al. [12], the effect of applied voltage on the friction coefficient was reversible and had nothing to do with the history of applied voltage. It was also believed that the applied electric field has ideal friction control performance for BH-1 ferrofluid, and the controllability of the friction coefficient was not a common property of lubricating materials.

5.1.3 Magnetic field control

Muju et al. [13] studied the influence of magnetic fields on wear and found that it was feasible to use a magnetic field to reduce the wear of nonmagnetic materials, and the relative hardness and temperature of the materials had an effect on wear. Yamamoto [14] found that the wear rate decreased after the external magnetic field was applied, and the effect of a DC magnetic field was better than that of an AC magnetic field. They thought that the surface activity of carbon steel and the resistance of the adsorption film to thermal disturbance were enhanced after the magnetic field was applied. The enhancement of boundary lubrication was not detected for nonpolar oil, but was for polar oil.

Ke et al. studied the migration of magnetohydrodynamic (MHD) controlled lubricants [15]. The results showed that temperature and external magnetic field strength had an influence on the MHD lubrication. Due to the influence of temperature gradient, the magnetic fluid migrated from the friction area, which took away the lubricant and caused a lack of oil. By controlling the intensity of the external magnetic field, a stable friction coefficient could be formed. Similar studies [16–18] showed that the existence of an external magnetic field was conducive to improving the friction and wear resistance of materials. The mechanism of action could be divided into two aspects: on the surface of the material and on the inside. The former used the magnetic field to attract the wear debris, promoted interface oxidation, and attracted the polar lubricant to prevent direct contact between the two surfaces of the friction pair; and the latter used magnetostriction and the dislocation movement to improve the wear resistance of the material.

5.1.4 Ultrasonic control

Research on the effect of ultrasonics on friction and wear of a friction pair can be traced back to the 1960s. Sase et al. [19] used an external shock wave to reduce the static friction coefficient of a guide rail. They believed that the larger the sliding speed, the smaller the friction coefficient; the stronger the shock wave energy and the smaller the slider mass, the smaller the friction coefficient. At present, research on ultrasonics and friction wear is mainly focused on the design of ultrasonic motors. Qu et al. [20] studied the mechanism of friction reduction by ultrasonic vibration. He believed that the ultrasonic vibration on the motor stator was decomposed into horizontal vibration and vertical vibration. The horizontal vibration produced a friction driving force, and the vertical ultrasonic vibration reduced the dynamic friction coefficient of ultrasonic driving. It was considered that the surface adhesion, plastic deformation, and ultrasonic amplitude were related, and the contact area shape also had a significant impact on the surface friction.

5.1.5 Application examples

Current collection through a brush is one example of this kind of active control technology. The brush is not only a friction pair but is also affected by the electric field through current. Watanabe et al. [21] studied the material ratio and wear of a space brush, and analyzed the influence of contact resistance, passing current, and material formula (mixture of silver and solid lubricant MoS_2, etc.) on the wear volume. Kohno et al. [22] studied the wear mechanism of the pure copper-steel friction pair of a current collector and found that discharge heating was one of the important reasons for wear.

A long-term study was conducted on the failure of rolling bearings under the action of an electric field or magnetic field. Wang [23] found that, under the action of an electric field, corrugations and pits formed in the inner and outer raceways of a rolling bearing, and the corrugations on the inner raceway were smaller than those on the outer raceway. This corrugation reduced the contact area and led to early failure of the bearing.

As for field-controlled lubrication, the electric field or magnetic field is mainly applied to control the lubrication characteristics of friction pairs. Charged vegetable lubricants were studied by Huang et al. [24]. LB-2000 vegetable lubricating oil was modified by the surfactant 12 alkylbenzene sulfonate, and a vegetable lubricant suitable for electrostatic spray lubrication was obtained. The results showed that the adsorption film and metal oxide film work together at the friction interface to promote the formation of lubricating film, which caused the charged vegetable oil to have good antifriction and antiwear properties. Zhou studied the spiral groove mechanical seal characteristics of magnetic fluid [25] and discussed the influence laws of magnetic field strength and rotation speed on oil film thickness, friction torque, characteristic temperature, consumption of magnetic fluid, etc. The results showed that the dynamic pressure of the fluid increases with the increase of magnetic field strength, and the appropriate sealing capacity could be obtained by adjusting the electric flow strength.

5.1.6 Conclusions

Using electric fields, magnetic fields, and ultrasonic fields to actively control the friction and wear process and state of a friction pair has practical engineering value for mechanical designs, and should be given more attention in machine design. The analysis of the effect of the external physical field on the adhesion wear, fatigue wear, and abrasive wear of the friction pair in the machine is of practical significance for the correct design of the machine and the prediction of the failure form and control of the machine, allowing active control of the failure of the friction pair in the machine and prolonging the service life of the machine through more correct design.

However, some problems remain to be solved. In particular, these measures need certain external field conditions, leading to a complex mechanical structure and difficulty in finding a magnetofluid to provide the function of stable directional lubrication over a long period. At the same time, other questions remain. How can the lubrication role of solid particles be accomplished with the coexistence of many solid particles? How can the function of soot particles in engine lubricating oil be controlled? Although many kinds of nanoparticles have antiwear and antifriction effects, how can the interaction between these nanoparticles and lubricant additives at the friction interface be controlled, to achieve the effect of collaborative lubrication and long-term lubrication? These problems need urgent solutions.

Using Fe_3O_4 nanoparticles as a magnetic core to carry solid lubricants to the rubbing surfaces has obvious advantages, such as mature synthesis technology and strong particle magnetism. However, if the Fe_3O_4 content is too low, the magnetic adsorption on the rubbing surfaces is insufficient, and if the Fe_3O_4 content is too high, the solid lubricant role cannot be played adequately, leading to greater wear and surface destruction, so this needs to be optimized in combination with the friction experiment results. Graphene is a friction reducing and wear-resistant material with strong adsorption capacity. Graphene with MoS_2 composite can have a synergistic lubrication effect. Therefore, using graphene oxide as the loading body of a core-shell microsphere is not only conducive to the dispersion of the nanoparticles, but also helpful in improving the lubrication performance of the microspheres. Thus it is useful to study the preparation and tribological performance of the magnetic particles (rGO@Fe_3O_4) in PAO base oil with or without biodiesel soot.

5.2 Characterization of BDS and rGO@Fe_3O_4

5.2.1 Introduction

Due to the extensive use of diesel engines, soot discharged from their combustion chambers is one of the primary environmental pollution sources [26, 27]. Not only is soot harmful to human health and ecological systems [28], it can also contaminate the lubricating oil in the sump as a consequence of blow-by gases [29]. This may increase the viscosity of the lubricating oil, decrease its lubricating properties, and reduce the working life of critical components in the engine [30–33]. Therefore, soot has a significant impact on engine parts and lubricating oil.

Due to its low soot emission, biodiesel has recently received considerable attention as a suitable diesel replacement [34]. Animal and vegetable oils, which are abundant worldwide, are the source of biodiesel. These substances can be converted into liquid fuels by transesterification. Biodiesel is preferable over petroleum diesel as it is environmentally friendlier, with lower pollution emissions, along with good combustion performance and biodegradability. Therefore, the conversion of animal and vegetable oils into biodiesel is considered to be one of the most promising approaches for energy conversion. However, biodiesel soot (BDS) can still contaminate the engine lubricating oil during use [35]. Certainly, the increase in biodiesel production will lead to its widespread use in engines [36, 37]; therefore, it is essential to analyze how the effects of BDS on lubrication performance of engine lubricating oil can be reduced.

Graphene has emerged as a promising lubrication material because of its layered structure and unique physicochemical properties [38, 39]. Wang et al. [40] reported that an amino-functionalized graphene reinforced PTFE composite showed good tribological performance under water-lubricated conditions. Gupta et al. [41] found that reduced graphene oxide (rGO) could effectively improve the friction-reducing and wear-resistance properties of poly(ethylene glycol). To further enhance the lubrication performance of graphene, many graphene/metal oxide composites have been proposed. Zhang et al. [42] and Wang et al. [43] observed that nanoparticles contribute to inhibiting the aggregation of graphene. Song et al. [44] found that the lubricating effect of α-Fe_2O_3 nanorod/graphene oxide (GO) was superior to that of only α-Fe_2O_3 or GO when used as additives in paraffin oil. Although previous studies have investigated the lubricating properties of graphene decorated with metal oxides, most have focused on pure lubricating oil. The effect of graphene decorated with metal oxides on the tribological behavior of soot-contaminated lubricating oil has been rarely reported. In this work, the tribological behavior of rGO@Fe_3O_4 as an additive in BDS-contaminated polyalphaolefin-6 (PAO) base oil was investigated. Furthermore, this section discusses the tribological mechanism. The results of this work may be helpful in optimizing the composition of additives in engine lubricant formulation to extend the life of engine components.

5.2.2 Materials and preparation

The rGO@Fe_3O_4 was acquired from NanoInnova Technologies, Spain. BDS was collected from the combustion of biodiesel by a lab-made soot-capture device. The detailed procedure for collecting the BDS was previously reported [45].

5.2.3 Characterization and analysis

The morphologies of BDS and rGO@Fe_3O_4 were analyzed with field-emission scanning electron microscopy (FESEM, SU8010, Hitachi, Japan) and field-emission transmission electron microscopy (FETEM, JEM-2100F, JEOL, Japan). The Raman spectra of BDS and rGO@Fe_3O_4 were also obtained (HR Evolution, Horiba Jobin Yvon, France).

5.2.4 BDS and rGO@Fe_3O_4 characterization

The morphology and chemical composition of BDS were analyzed by SEM, TEM, and Raman spectroscopy (Fig. 5.1). Agglomerated BDS particles with irregular shapes can be observed in Fig. 5.1A [46], and the size of the agglomerate was approximately 10–25 μm [47]. Fig. 5.1B and C show that most of the primary BDS particles form chain-like aggregates. Moreover, they have a nearly spherical shape and are composed of amorphous phases and graphitic layers. There was a direct correlation between the aggregation of BDS primary particles and the increase of engine wear [32, 48]. The Raman spectrum shows the disorder (D) band and graphitic (G) bands at 1345 and 1590/cm, indicating that BDS is composed of carbonaceous materials [49].

The morphology and chemical composition of rGO@Fe_3O_4 are observed in Fig. 5.2. The FESEM images in Fig. 5.2A display rGO sheets with Fe_3O_4 nanoparticles distributed on them. As shown in Fig. 5.2B and C, despite the many wrinkles on the rGO sheets, Fe_3O_4 nanoparticles were evenly dispersed [50]. Moreover, Fe_3O_4 nanoparticles had a nearly spherical

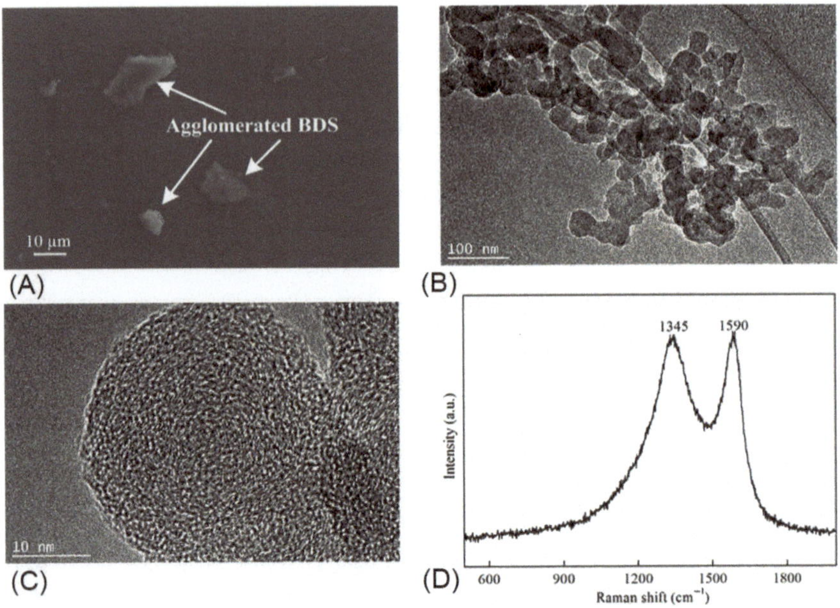

Fig. 5.1 (A) FESEM and (B and C) FETEM images and (D) Raman spectrum of BDS.

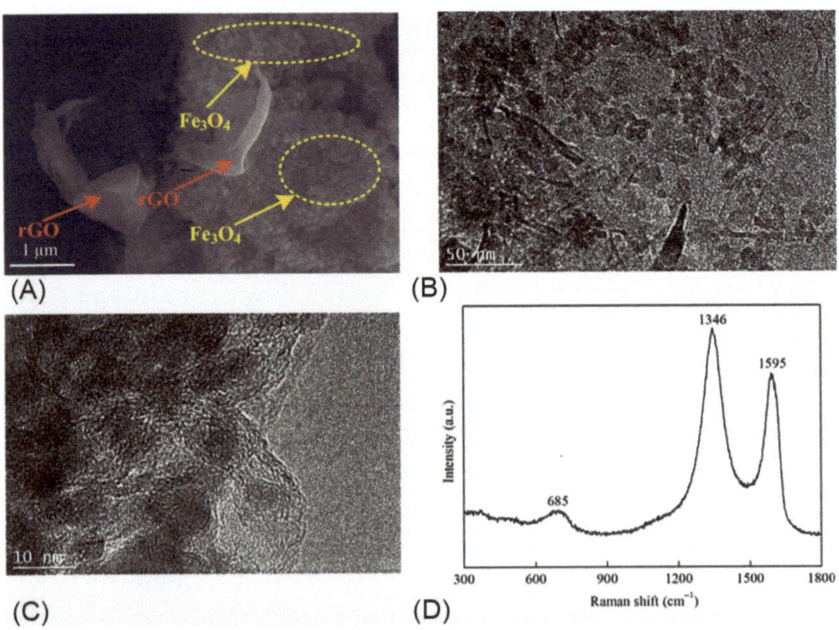

Fig. 5.2 (A) FESEM and (B and C) FETEM images and (D) Raman spectrum of rGO@Fe_3O_4.

shape with a diameter of approximately 10 nm. The Raman spectrum shows bands at 685, 1346, and 1595/cm that can be ascribed to the Fe_3O_4 nanoparticles, and D and G bands of rGO, indicating that Fe_3O_4 nanoparticles are found on the surface of the rGO sheets.

5.3 Effect of rGO@Fe_3O_4 on the tribological behavior of biodiesel soot-contaminated polyalphaolefin

5.3.1 Introduction

Zhang et al. [42] reported that reduced graphene oxide/Fe_3O_4 (RGO/Fe_3O_4) composites were fabricated by depositing Fe_3O_4 nanoparticles onto the surface of RGO. After modification, RGO/Fe_3O_4 was dispersed in low-viscosity PAO6. The tribological properties of RGO/Fe_3O_4 in low-viscosity PAO6 were evaluated using a CETR-UMT tribometer. The wear mechanism of RGO/Fe_3O_4 was also discussed. The results showed that base oil with 0.1 wt% RGO/Fe_3O_4 showed extreme antiwear performance, with the average wear volume of the steel disk decreased by 52.27% compared to base oil. This could be attributed to the emergence of a positive synergistic effect between Fe_3O_4 and GO, which promotes the formation of a tribofilm. The results of the Raman spectra and XPS proved that RGO/Fe_3O_4 in base oil participates in the formation of a tribofilm on worn surface. Therefore, the addition of RGO/Fe_3O_4 into PAO6 base oil can significantly improve the antiwear performance of the oil.

5.3.2 Materials and preparation

PAO6 base oil was purchased from Chevron Phillips Chemical Co., United States. Acetone was purchased from Sinopharm Chemical Reagent Co., Ltd., China. The oil samples were prepared as follows. PAO6 base oil with 2 wt% and 5 wt% BDS were selected as the lower and upper soot contents [31, 46]. For this, BDS was added into PAO6 base oil at 0, 2, and 5 wt%, and then distributed using a glass rod for 15 min and mixed for 1 h with a magnetic stirrer. Subsequently, rGO@Fe_3O_4 was added into the mixtures (PAO6, PAO6+2 wt% BDS, and PAO6+5 wt% BDS) at 0.25 wt%, 0.5 wt%, 1 wt%, and 2 wt% to obtain a total of 12 samples. The products were then ultrasonically treated for 30 min.

5.3.3 Tribological tests

Tribological experiments were performed using a UMT-2 multifunction friction tester (Bruker Co., Ltd, United States). The upper ball had a 6-mm

Fig. 5.3 Schematic illustration of the tribological test and analysis methods.

diameter with a 647–861 HV hardness, and the lower disk had a 10-mm diameter × 3-mm height with 201–245 HV in hardness. The friction pairs were made of GCr15 steel. A schematic illustration of the tribological test is shown in Fig. 5.3. A reciprocating sliding ball-on-disk system was used. All of the experiments were performed under a 10 N load for 30 min with a sliding speed and amplitude of 10 mm/s and 5 mm. During the tribological test, the friction coefficient was automatically measured by a computer connected to the tribotesting machine. Each friction test was repeated three times.

5.3.4 Characterization and analysis

Wear volume, three-dimensional images, and the corresponding worn surface profiles of the wear scar of the lower disc were measured by three-dimensional laser scanning microscopy (VK-X100, Keyence, Japan). The morphology and chemical composition of the worn surfaces of the lower disc were analyzed by FESEM and Raman spectroscopy.

5.3.5 Friction reduction and wear resistance

Fig. 5.4 shows the effect of rGO@Fe_3O_4 content on the average friction coefficient and wear volume of PAO6 base oil and PAO6 base oil with 2 wt% BDS. Overall, the average friction coefficient decreased with increasing rGO@Fe_3O_4 content and then increased; see Fig. 5.4A. When comparing

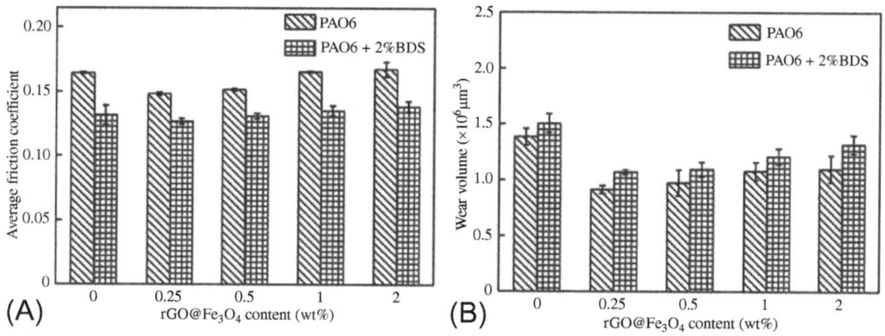

Fig. 5.4 Effect of $rGO@Fe_3O_4$ content on (A) average friction coefficient and (B) wear volume of pure PAO6 base oil and PAO6 base oil with 2 wt% BDS.

samples with the same $rGO@Fe_3O_4$ content, the average friction coefficient of PAO6 base oil with 2 wt% BDS was lower than that of pure PAO6 base oil. Both samples, with and without BDS, exhibited the lowest average friction coefficient when the content of $rGO@Fe_3O_4$ was 0.25 wt%. Specifically, the average friction coefficient of PAO6 base oil was reduced from 0.165 without $rGO@Fe_3O_4$ to 0.148 with 0.25 wt% $rGO@Fe_3O_4$, and that of PAO6 base oil with 2 wt% BDS was reduced from 0.132 to 0.127. These results indicate that BDS could reduce the friction of PAO6 base oil [29]. Moreover, $rGO@Fe_3O_4$ could further enhance the antifriction property of PAO6 base oil and PAO6 base oil with 2 wt% BDS. However, when the $rGO@Fe_3O_4$ content was too high, both PAO6 base oil and PAO6 base oil with 2 wt% BDS showed poor performance in friction reduction, which may be ascribed to the aggregation of $rGO@Fe_3O_4$ [51].

Fig. 5.4B shows that with the addition of $rGO@Fe_3O_4$ to PAO6 base oil and PAO6 base oil with 2 wt% BDS, the wear volume significantly decreased at first, and then increased. The best wear-resistance performance was for 0.25 wt% of $rGO@Fe_3O_4$: the wear volume of PAO6 base oil was reduced by 34.2% (from 1.39×10^6 to 9.15×10^5 μm^3) and that of the oil with 2 wt% BDS was reduced by 29.1% (from 1.51×10^6 to 1.07×10^6 μm^3). This confirms that $rGO@Fe_3O_4$ effectively improved the antiwear properties of PAO6 base oil and PAO6 base oil with 2 wt% BDS.

The effects of $rGO@Fe_3O_4$ content on the average friction coefficient and wear volume of PAO6 base oil and PAO6 base oil with 5 wt% BDS are shown in Fig. 5.5. As seen in Fig. 5.5A, when adding $rGO@Fe_3O_4$ particles to PAO6 base oil with 5 wt% BDS, the average friction coefficient initially decreased and then increased, similar to the PAO6 base oil with 2 wt% BDS. This indicates that $rGO@Fe_3O_4$ could also slightly improve the antifriction

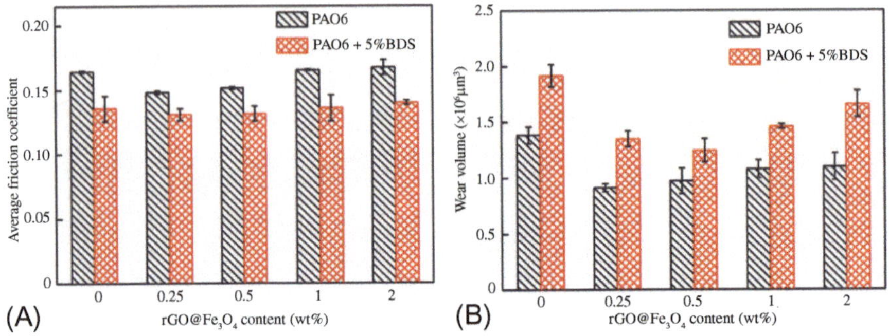

Fig. 5.5 Effect of rGO@Fe_3O_4 content on (A) friction coefficient and (B) wear volume of pure PAO6 base oil and PAO6 base oil with 5 wt% BDS.

characteristics of PAO6 base oil with 5 wt% BDS. The wear volume of PAO6 with 5 wt% BDS and 0.5 wt% rGO@Fe_3O_4 reached a minimum (1.25×10^6 μm^3), which is 34.9% lower than the sample with no additive (1.92×10^6 μm^3); see Fig. 5.5B. However, when the rGO@Fe_3O_4 content exceeded 0.5 wt%, the wear volume gradually increased.

5.3.6 Analysis of tribological mechanism

Three-dimensional images and the corresponding worn surface profiles of the wear scar of the lower disc lubricated with different samples are shown in Fig. 5.6. There are remarkable differences in the degree of wear in each case. According to Fig. 5.6A–C and A'–C', the wear area on the lower disc lubricated with PAO6 base oil with 5 wt% BDS was the largest, followed by that of the disc lubricated with PAO6 base oil containing 2 wt% BDS; the smallest was that lubricated with PAO6 base oil, i.e., the abrasion wear increased with the increase in BDS content [52, 53]. Compared to that of PAO6 base oil, the wear scar of the lower disc lubricated with PAO6 base oil containing 0.25 wt% rGO@Fe_3O_4 was shallow and narrow; see Fig. 5.6D and d′. When 0.25 wt% rGO@Fe_3O_4 was added to PAO6 base oil with 2 wt% BDS, the wear area was remarkably reduced, as seen in Fig. 5.6E and E′. The addition of 0.5 wt% rGO@Fe_3O_4 led to a significant decrease in the wear degree of the disc lubricated with PAO6 5 wt% BDS, shown in Fig. 5.6F and F′. These results confirm that rGO@Fe_3O_4 improves the tribological properties of pure and BDS-containing PAO base oil.

FESEM images and Raman spectra of the worn surfaces of the lower disc are shown in Fig. 5.7. The relative wear degree observed in the FESEM images was divided into shallow, moderate, and severe [54]. Moderate furrows were observed on the worn surfaces of the lower disc lubricated with

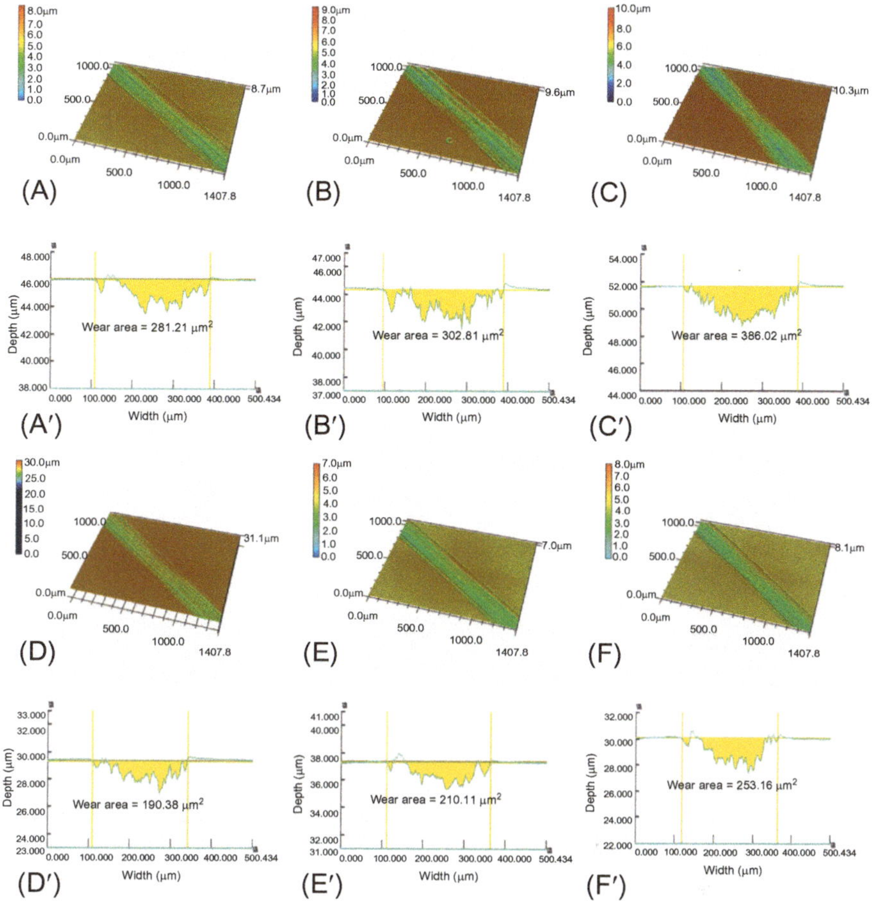

Fig. 5.6 **Three-dimensional images and the corresponding worn surface profiles of the wear scar in the lower disc lubricated with:** (A and A′) PAO6, (B and B′) PAO6 + 2 wt% BDS, (C and C′) PAO6 + 5 wt% BDS, (D and D′) PAO6 + 0.25 wt% rGO@Fe_3O_4, (E and E′) PAO6 + 2 wt% BDS + 0.25wt% rGO@Fe_3O_4 and (F and F′) PAO6 + 5 wt% BDS + 0.5 wt% rGO@Fe_3O_4.

PAO6 base oil, as seen in Fig. 5.7A. The corresponding Raman spectrum, in Fig. 5.7A′, shows weak D and G bands at 1349 and 1572/cm, which are attributed to the carbonization of PAO6 base oil during friction [55]. In addition, the bands positioned at 539 and 667/cm can be ascribed to Fe_3O_4, which may be due to the tribochemical production of steel [56].

For the surfaces lubricated with PAO6 base oil with 2 wt% BDS, in Fig. 5.7B, the furrows are severe compared to those of the surface lubricated with PAO6 base oil, which presented moderate furrows. Moreover, the highest wear degree was for the surface lubricated with PAO6 base oil

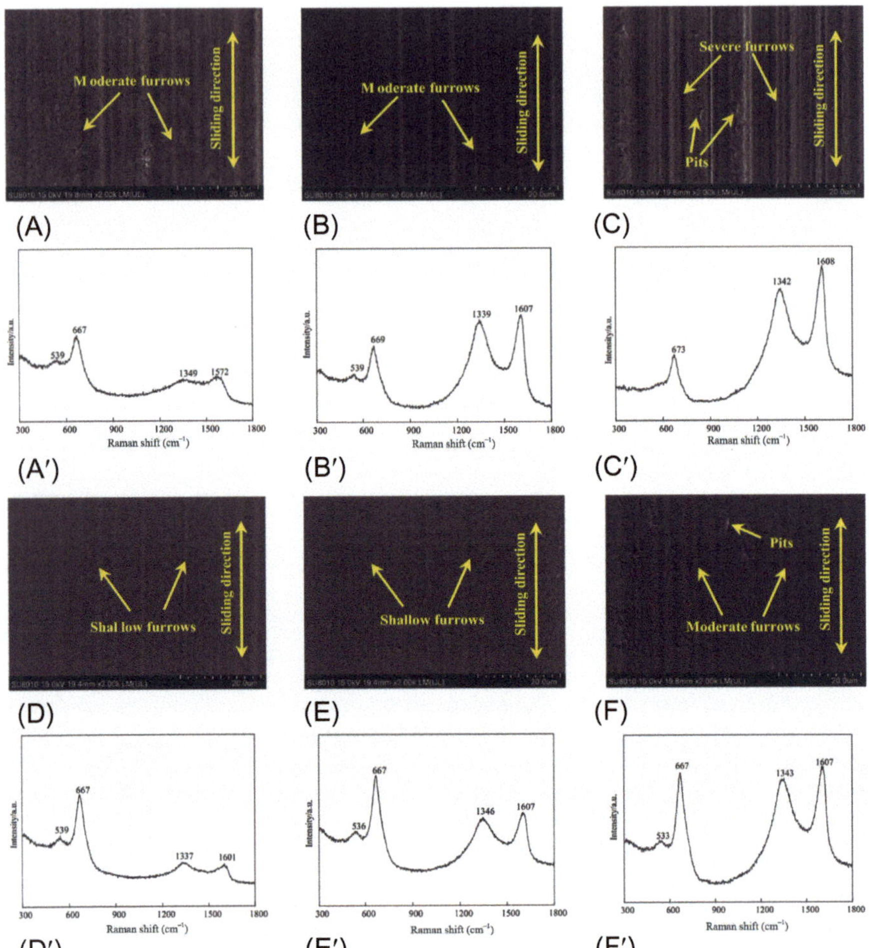

Fig. 5.7 FESEM images and Raman spectra of the worn surfaces of the lower disc lubricated with: (A and A′) PAO6, (B and B′) PAO6 + 2 wt% BDS, (C and C′) PAO6 + 5 wt% BDS, (D and D′) PAO6 + 0.25 wt% rGO@Fe_3O_4, (E and E′) PAO6 + 2 wt% BDS + 0.25 wt% rGO@Fe_3O_4 and (F and F′) PAO6 + 5 wt% BDS + 0.5 wt% rGO@Fe_3O_4.

5 wt% BDS, in which many remarkably severe furrows and some wear pits can be observed; see Fig. 5.7C. As shown in Fig. 5.7B′ and C′, the intensity of the D and G bands at 1339 and 1607/cm of the worn surfaces lubricated with PAO6 base oil 2 wt% BDS increased compared to those using pure PAO6 base oil for lubrication. Moreover, the intensity of these bands (1342 and 1609/cm) was even stronger when lubricating with PAO6 base oil 5 wt% BDS. The results indicate that the larger BDS content led to an increase in BDS adsorption on the friction surfaces, inducing a higher wear [35, 57].

As seen from Fig. 5.7D–F, the furrows of the worn surface significantly decreased when 0.25 wt% rGO@Fe_3O_4 was added to PAO6 base oil and PAO6 base oil with 2 wt% BDS; their wear degrees were shallow. In addition, there were some light furrows and a few pits on the worn surfaces lubricated with PAO6 base oil containing 5 wt% BDS and 0.5 wt% rGO@Fe_3O_4, which can be described as moderate wear. Thus rGO@Fe_3O_4 greatly improves the performance of the three lubricants. This agrees with the experimental three-dimensional laser scanning microscopy analysis. The Raman spectra in Fig. 5.7D′–F′, show D and G bands at 1337–1343/cm and 1601–1607/cm. The bands at 533–539/cm and 667/cm correspond to Fe_3O_4. It is worth noting that the intensity of the D and G bands from BDS are much stronger than those from rGO@Fe_3O_4, because the BDS content (2 wt% or 5 wt%) is higher than that of the additive (0.25 wt% or 0.5 wt%). Therefore, the Raman spectra on the worn surface only show the BDS bands and those of Fe_3O_4 in Fig. 5.7E′ and F′. This is consistent with the results reported by Su et al. [58] Nevertheless, the intensity of the Fe_3O_4 band of the worn surfaces increased when rGO@Fe_3O_4 was added to the BDS-containing lubricant. This indicates that there is a close relationship between the improvement of antiwear properties and the Fe_3O_4 content, which also implies that rGO@Fe_3O_4 participates in the formation of the tribofilm during the friction process [59, 60]. Moreover, during friction, part of the Fe_3O_4 particles on the surface of rGO may be exfoliated due to friction and shear force. These particles could fill the friction contact area and repair the worn surface. This result agrees with the work by Zhang et al. [42] and Huang et al. [51].

Therefore, when BDS exists in the lubricating oil, it may directly abrade the tribofilm and the surface of the engine parts, leading to increased wear [61]. With the addition of rGO@Fe_3O_4 in PAO6 base oil with BDS, the antiwear properties could be significantly improved. The corresponding tribological mechanism may be as follows. First, rGO@Fe_3O_4 is adsorbed on the friction interface to prevent direct contact of friction pairs, and between the BDS and friction pair surface because of the two-dimensional structure and unique extremely high strength of rGO [62–65]. Then, during the sliding motion, the exfoliated Fe_3O_4 particles from rGO@Fe_3O_4 can enter into the frictional interfaces and could induce the self-repairing effect.

5.3.7 Conclusions

In summary, a UMT-2 multifunction friction tester was used to investigate the effects of rGO@Fe_3O_4 on the tribological behavior of BDS-contaminated

PAO6 base oil. Adding rGO@Fe_3O_4 can effectively improve the antiwear properties of PAO6 base oil with BDS. When the content of rGO@Fe_3O_4 additive was 0.25 wt%, the wear volume of PAO6 base oil with 2 wt% BDS was reduced by 29.1% (from 1.51×10^6 to 1.07×10^6 μm^3). With 0.5 wt% rGO@Fe_3O_4, the wear volume of PAO6 base oil with 5 wt% BDS decreased by 34.9% (from 1.92×10^6 to 1.25×10^6 μm^3). The decreased wear is attributed to the formation of a protecting layer in the presence of rGO@Fe_3O_4 and the shifting of exfoliated Fe_3O_4 particles into the contact region to repair the worn surfaces.

References

[1] Nian J, Si Y, Guo Z. Advances in atomic-scale tribological mechanisms of solid interfaces. Tribol Int 2016;94:1–13.
[2] Xu J, Li J. New achievements in superlubricity from International Workshop on Superlubricity: fundamental and applications. Friction 2015;3(4):344–51.
[3] Konicek AR, Jacobs PW, Webster MN, Schilowitz AM. Role of tribofilms in wear protection. Tribol Int 2016;94:14–9.
[4] Thakre AA, Shinde A, Mundhe G. Improvement in boundary lubrication characteristics of SAE20W40 oil using aluminum oxide nanoparticles. J Tribol 2016;138(3), 034501.
[5] Marko M, Kyle JP, Wang Y, Branson B, Terrell EJ. Numerical and experimental tribological investigations of diamond nanoparticles. J Tribol 2016;138(3), 032001.
[6] Wu Y, Tsui WC, Liu TC. Experimental analysis of tribological properties of lubricating oils with nanoparticle additives. Wear 2007;262(7):819–25.
[7] Wang J, Xue Z. Research progress of active control in friction and wear. Sci Technol Innov Herald 2007;(33):41–3.
[8] Hurricks PL. The effect of applied voltage on the frictional behavior of carbon black filled elastomers. Wear 1979;52:365–80.
[9] Yamamoto Y, Hirano F. Scuffing resistance of phosphate esters II: effect of applied voltage. Wear 1981;66:77–86.
[10] Iwai K, Yamamoto Y, Takahashi S. Relationship between tribology characteristics and induced-voltage magnetic field in ring-rod-type dry wear test. Proc JSME Part C 1995;61(587):3085–92.
[11] Zhai W, Yamamoto Y. The effect of externally applied voltage on tribological characteristics of steel rubbing pairs under boundary lubrication. Tribology 2000;20(6): 435–8.
[12] Jiang HJ, Meng YG, Wen SZ. Active control of friction by applying electric fields across boundary films of stearate. Proc ASIATRIB 1998;755–60.
[13] Muju MK, Ghosh A. Effect of a magnetic field on the diffusive wear of cutting tools. Wear 1980;58:137–45.
[14] Yamamoto Y, Gondo S. Effect of a magnetic field on boundary lubrication. Tribol Int 1987;20(6):342–6.
[15] Ke H, Huang W, Wang X. Controlling lubricant migration using ferrofluids. Tribol Int 2016;93 A:318–23.
[16] Wei Y, Zhang Y, Chen Y. Study on the magnetic field interference mechanism of the dry tribological characteristics of the zinc brass aluminum alloy pair. Tribology 2013;33(2):184–90.

[17] Wang H, Yan L, Liu D, Song H. Investigation of the tribological properties: core-shell structured magnetic Ni@NiO nanoparticles reinforced epoxy nanocomposites. Tribol Int 2015;83:139–45.
[18] Hsu TC, Chen JH, Chiang HL, Chou T. Lubrication performance of short journal bearings considering the effects of surface roughness and magnetic field. Tribol Int 2013;61:169–75.
[19] Sase N, Okano R, Chandy M, et al. Control of guide way friction by stress wave-effect of a single impact on the reduction in friction. Proc JSME Part C 1993;59(567):3488–93.
[20] Luo Y, Qu J, Qi Y. Antifriction mechanism of ultrasonic vibration based on ultrasonic micro driving. Chin J Mech Eng 2001;37(5):52–6.
[21] Watanabe M, Suzuki M, Nishlmura M. Wear and electric characteristics of slip-ring materials for space applications, part 1—influence of the load. J Jpn Soc Tribol 1995;41(5):419–26.
[22] Kohno A, Ohtake M, Mishina H, Kanekama U. Wear mechanism of pure copper and steel in current collecting system. J Jpn Soc Tribol 1996;42(4):312–5.
[23] Wang J. Effect of electrical and magnetic field on the bearing life. Bear 2003;8:38–41.
[24] Huang S, Li Z, Yao W. Tribological properties of charged vegetable oil. Tribology 2014;34(4):371–8.
[25] Zhou J, Fan H, Shao C. Experimental study on the hydrodynamic lubrication characteristics of magnetofluid film in a spiral groove mechanical seal. Tribol Int 2016;95:192–8.
[26] Geng P, Cao EM, Tan QM, Wei LJ. Effects of alternative fuels on the combustion characteristics and emission products from diesel engines: a review. Renew Sustain Energy Rev 2017;71:523–34.
[27] Alptekin E. Emission, injection and combustion characteristics of biodiesel and oxygenated fuel blends in a common rail diesel engine. Energy 2017;119:44–52.
[28] Shahir VK, Jawahar CP, Suresh PR. Comparative study of diesel and biodiesel on CI engine with emphasis to emissions—a review. Renew Sustain Energy Rev 2015;45:686–97.
[29] Hu EZ, Hu XG, Liu TX, Fang L, Dearn KD, Xu HM. The role of soot particles in the tribological behavior of engine lubricating oils. Wear 2013;304:152–61.
[30] Green DA, Lewis R, Dwyer-Joyce RS. Wear effects and mechanisms of soot-contaminated automotive lubricants. Proc Inst Mech Eng Part J J Eng Tribol 2006;220:159–69.
[31] Penchaliah R, Harvey TJ, Wood RJK, Nelson K, Powrie HEG. The effects of diesel contaminants on tribological performance on sliding steel on steel contacts. Proc Inst Mech Eng Part J J Eng Tribol 2011;225:779–97.
[32] Esangbedo C, Boehman AL, Perez JM. Characteristics of diesel engine soot that lead to excessive oil thickening. Tribol Int 2012;47:194–203.
[33] Kontou A, Southby M, Morgan N, Spikes HA. Influence of dispersant and ZDDP on soot wear. Tribol Lett 2018;66:157.
[34] Liu HF, Bi XJ, Huo M, Lee CFF, Yao M. Soot emissions of various oxygenated biofuels in conventional diesel combustion and low-temperature combustion conditions. Energy Fuel 2012;26:1900–11.
[35] Li C, Wei DZ, Zhang Y, Song RH, Hu XG. Effect of biodiesel soot on tribological behavior of liquid paraffin. Chin Pet Process Pe Technol 2018;20:106–13.
[36] Hajjari M, Tabatabaei M, Aghbashlo M, Ghanavati H. A review on the prospects of sustainable biodiesel production: a global scenario with an emphasis on waste-oil biodiesel utilization. Renew Sustain Energy Rev 2017;72:445–64.
[37] Monirul IM, Masjuki HH, Kalam MA, Mosarof MH, Zulkifli NWM, Teoh YH, How HG. Assessment of performance, emission and combustion characteristics of palm, jatropha and *Calophyllum inophyllum* biodiesel blends. Fuel 2016;181:985–95.

[38] Zheng D, Cai ZB, Shen MX, Li ZY, Zhu MH. Investigation of the tribology behaviour of the graphene nanosheets as oil additives on textured alloy cast iron surface. Appl Surf Sci 2016;387:66–75.
[39] Zhang GQ, Xu Y, Xiang XZ, Zheng GL, Zeng XQ, Li ZP, Ren TH, Zhang YD. Tribological performances of highly dispersed graphene oxide derivatives in vegetable oil. Tribol Int 2018;126:39–48.
[40] Wang X, Wu JW, Zhou LH, Wei XC, Wang WR. Tribological behaviors of amino functionalized graphene reinforced PTFE composite. Proc Inst Mech Eng Part J J Eng Tribol 2018;232:1428–36.
[41] Gupta B, Kumar N, Panda K, Melvin AA, Joshi S, Dash S, Tyagi AK. Effective noncovalent functionalization of poly(ethylene glycol) to reduced graphene oxide nanosheets through γ-radiolysis for enhanced lubrication. J Phys Chem C 2016;120:2139–48.
[42] Zhang QQ, Wu B, Song RH, Song H, Zhang J, Hu XG. Preparation, characterization and tribological properties of polyalphaolefin with magnetic reduced graphene oxide/Fe_3O_4. Tribol Int 2020;141:105952.
[43] Wang Y, Zhang S, Chen H, Li HX, Zhang P, Zhang ZY, Liang GH, Kong JL. One-pot facile decoration of graphene nanosheets with Ag nanoparticles for electrochemical oxidation of methanol in alkaline solution. Electrochem Commun 2012;17:63–6.
[44] Song HJ, Jia XH, Li N, Yang XF, Tang H. Synthesis of α-Fe_2O_3 nanorod/graphene oxide composites and their tribological properties. J Mater Chem 2012;22:895–902.
[45] Li C, Li ML, Wang XY, Feng WM, Zhang QQ, Wu B, Hu XG. Novel carbon nanoparticles derived from biodiesel soot as lubricant additives. Nanomater 2019;9:1115.
[46] Salehi FM, Khaemba DN, Morina A, Neville A. Corrosive–abrasive wear induced by soot in boundary lubrication regime. Tribol Lett 2016;63:19.
[47] Hu EZ, Hu XG, Liu TX, Liu YM, Song RH, Chen YZ. Investigation of morphology, structure and composition of biomass-oil soot particles. Appl Surf Sci 2013;270:596–603.
[48] La Rocca A, Di Liberto G, Shayler PJ, Fay MW. The nanostructure of soot-in-oil particles and agglomerates from an automotive diesel engine. Tribol Int 2013;61:80–7.
[49] Patel M, Ricardo CLA, Scardi P, Aswath PB. Morphology, structure and chemistry of extracted diesel soot—part I: transmission electron microscopy, Raman spectroscopy, X-ray photoelectron spectroscopy and synchrotron X-ray diffraction study. Tribol Int 2012;52:29–39.
[50] Zong M, Huang Y, Zhao Y, Wang L, Liu PB, Wang Y, Wang QF. One-pot simplified co-precipitation synthesis of reduced graphene oxide/Fe_3O_4 composite and its microwave electromagnetic properties. Mater Lett 2013;106:22–5.
[51] Huang J, Li Y, Jia XH, Song HJ. Preparation and tribological properties of core-shell Fe_3O_4@C microspheres. Tribol Int 2019;129:427–35.
[52] George S, Balla S, Gautam M. Effect of diesel soot contaminated oil on engine wear. Wear 2007;262:1113–22.
[53] Salehi FM, Morina A, Neville A. The effect of soot and diesel contamination on wear and friction of engine oil pump. Tribol Int 2017;115:285–96.
[54] Lim CYH, Lim SC, Gupta M. Wear behaviour of SiCp-reinforced magnesium matrix composites. Wear 2003;255:629–37.
[55] Shi B, Guo JH, Cao XA, Hu EZ, Hu KH. Effects of carbon soot from the combustion of diesel fuels on the tribological properties of lubricating oil and diesel fuels. Ind Lubr Tribol 2018;70:532–7.
[56] Zhang X, Niu YG, Meng XD, Li Y, Zhao JP. Structural evolution and characteristics of the phase transformations between α-Fe_2O_3, Fe_3O_4 and γ-Fe_2O_3 nanoparticles under reducing and oxidizing atmospheres. Cryst Eng Comm 2013;15:8166–72.
[57] Kontou A, Southby M, Spikes HA. Effect of steel hardness on soot wear. Wear 2017;390:236–45.

[58] Su FH, Chen GF, Sun JJ. Synthesis of hydrogenated DLC film by PECVD and its tribocorrosion behaviors under the lubricating condition of graphene oxide dispersed in water. Tribol Int 2019;130:1–8.
[59] Zhou GH, Zhu YF, Wang XM, Xia MJ, Zhang Y, Ding HY. Sliding tribological properties of 0.45% carbon steel lubricated with Fe_3O_4 magnetic nano-particle additives in base oil. Wear 2013;301:753–7.
[60] Xu YF, Geng J, Peng YB, Liu ZC, Yu JY, Hu XG. Lubricating mechanism of Fe_3O_4@MoS_2 core-shell nanocomposites as oil additives for steel/steel contact. Tribol Int 2018;121:241–51.
[61] Green DA, Lewis R. Effect of soot on oil properties and wear of engine components. J Phys D Appl Phys 2007;40:5488.
[62] Huang JG, Tan J, Fang H, Gong F, Wang J. Tribological and wear performances of graphene-oil nanofluid under industrial high-speed rotation. Tribol Int 2019;135:112–20.
[63] He AS, Huang SQ, Yun JH, Jiang ZY, Stokes JR, Jiao SH, Wang LZ. Tribological characteristics of aqueous graphene oxide, graphitic carbon nitride, and their mixed suspensions. Tribol Lett 2018;66:42.
[64] Li C, Zhang QQ, Wu B, Feng WM, Song H, Hu XG. Effect of rGO@Fe_3O_4 on the tribological behavior of biodiesel soot-contaminated polyalphaolefin. Proc Inst Mech Eng Part J J Eng Tribol 2020. Submitted.
[65] Li YR, Zhao J, Tang C, He YY, Wang YF, Chen J, Mao JY, Zhou QQ, Wang BY, Wei F, Luo JB, Shi GQ. Highly exfoliated reduced graphite oxide powders as efficient lubricant oil additives. Adv Mater Interfaces 2016;3:1600700.

CHAPTER 6

Functional transformation of biodiesel soot

6.1 Novel approach for improved tribological behavior of biodiesel soot in liquid paraffin

6.1.1 Introduction

As concerns regarding the continuous consumption of fossil fuels and global awareness of energy savings and the reduction of emissions increase, scientists worldwide are striving to develop new cleaner fuels. Biodiesel is one of the most promising alternative fuels for automotive diesel engines. Biodiesel is a fatty acid alkyl ester, which is obtained through the transesterification of animal fat, vegetable oils, or waste cooking oil with methanol [1]. Compared with petrochemical diesel, the sulfur and aromatics in biodiesel are lower, because the cleaner properties greatly reduce the content of sulfides and carcinogens in the exhaust gas and reduce the amount of pollution added to the environment. In addition, biodiesel contains 10 wt% to 11 wt% of oxygen, enhancing the rate of heat release during the combustion process and reducing the amount of emissions (CO, HC, and soot) [2]. However, a portion of the biodiesel soot (BDS) generated during combustion can still contaminate the engine oil within the sump as a result of the blow-by gases [3].

With the extensive application of exhaust gas recirculation systems (EGRs) in engines, the amount of soot in the engine oil is increased, which leads to an increase in the viscosity of the engine oil and an aggravation of the wear. Su et al. [4] reported that a high concentration of biodiesel soot in lubricating oil breaks the lubrication film and blocks the engine oil entering into the frictional region, resulting in a reduction of the tribological properties of the engine oil. However, Guo et al. [5] showed that the low concentration of soot (0.01 wt%) as lubricant additive in lubricating oil not only does not aggravate the wear, but rather it displays antiwear and friction-reduction properties. Moreover, our previous research [6] found

Biodiesel Soot
https://doi.org/10.1016/B978-0-12-820553-2.00006-2

that the graphitic layers in BDS are not continuous and are surrounded by amorphous phases. Merchan-Merchan and Lapuerta et al. [7, 8] revealed the presence of highly curved graphene segments in BDS, and showed that BDS displays more ordered graphite-like structures and lower amorphous carbon concentration than diesel soot. Wei et al. [9, 10] reported that thermally oxidized candle soot and acidified candle soot have a higher order than naked candle soot. The candle soot was heat-treated and modified using a cationic surfactant, which can obviously improve the lubricating properties of the liquid paraffin. Antusch et al. [11] analyzed the tribological behavior of different soot particles by means of a pin-on-disk tribometer. They found that the wear properties of the soot particles are closely related to their reactivity and the number of defect sites. Hirata et al. [12] found that defects in the outermost layer of the carbon onions increase the bonding with the counter materials, resulting in inferior tribological properties. Therefore, the higher the order of the BDS, the better the improved BDS lubricity achieved. However, there have been few reports on improving the tribological behavior of BDS in lubricating oil by increasing its order.

In the study described in this chapter, the order of BDS was increased through thermally oxidized treatment, improving the oil solubility through a modification with oleylamine (OLA). The tribological behaviors of BDS (before and after thermally oxidized oleylamine modified) in liquid paraffin (LP) were evaluated using a ball-on-disc reciprocating tribometer, and the corresponding tribological mechanisms were investigated. The aim of this work is to provide a basis for inhibiting the effects of BDS on the tribological properties of lubricating oil, and to lay a foundation for the use of BDS as solid lubricant additives.

6.1.2 Experiments

6.1.2.1 Materials and preparation

LP was purchased from Shanghai Zhongqin Chemical Reagent Co., Ltd. Oleylamine was purchased from Aladdin Industrial Co., Ltd. All of the chemicals used in the present work were of an analytical grade reagent. Biodiesel was produced through the transesterification of soybean oil with methanol in the presence of a self-made green biomass ash catalyst [13]. The physicochemical properties of biodiesel are shown in Table 6.1. BDS was obtained from the combustion of biodiesel using a self-made soot-capture device [6]. A brief description of the BDS collection is given here [14]. The biodiesel was burned in a porcelain crucible at normal temperature and pressure. The BDS was collected using a glass slide (BDS capture slide) placed above the biodiesel flame. As the biodiesel burned, the flame

Table 6.1 Physical properties of biodiesel.

Property	Value	Methods
Density at 20°C (kg/m^3)	875	ASTM D4052
Flash point (closed cup) (°C)	168	ASTM D93
Acid value (mg KOH/g)	0.045	ASTM D664
Kinematic viscosity at 40°C (mm^2/s)	4.5	ASTM D445
Water content (%)	Trace	ASTM D1744

deposited BDS on the slide. When a BDS layer of a few micrometers thick had been deposited, this layer was then very carefully scraped off the slide and collected in a glass bottle. Finally, the collected BDS was dried under vacuum conditions at 120°C for 3 h and ground to a fine granular powder. Then 200 mg of BDS was placed in a porcelain burning boat within a tube furnace (OTF-1200X, Hefei Kejing Materials Technology Co., Ltd.) at 500°C for 2 h under an air flow of 40 mL/min and then cooled at room temperature. This thermally oxidized BDS is denoted as T-BDS. Then, 100 mg of T-BDS was refluxed in 5 g of oleylamine at 110°C for 12 h. The mixture was purified through repeated washing with ethanol using centrifugation. Finally, the product was dried under vacuum conditions at 120°C for 5 h. The T-BDS modified with oleylamine is denoted as T-BDS-OLA. Two kinds of soot (BDS and T-BDS-OLA) were added into the LP at mass percentages of 0.1, 0.2, 0.3, and 0.4 wt%, separately. The mixtures were then distributed using a glass rod for 15 min, followed by magnetic stirring for 1 h to reduce deviations during the experiment.

6.1.2.2 Tribological tests and characterization

Tribological tests were carried out on a ball-on-disc reciprocating tribometer (CFT-I, ZKKH), as shown in Fig. 6.1. The upper ball sample (6ϕ mm) and lower plate sample (50 × 50 × 5 mm) were made of GCr15 steel with a hardness of 62-65 HRC. During the frictional experiment, the sliding distance was 5 mm. Eqs. (6.1) and (6.2) give the wear volume of the upper ball sample (V_{wear}) [15].

$$V_{wear} = \frac{\pi h}{6}\left(h^2 + 3r^2\right), \tag{6.1}$$

$$h = R - \sqrt{R^2 - r^2}, \tag{6.2}$$

where r is the radius of the wear scar of the upper ball, h is the height of the wear of the upper ball, and R is the radius of the upper ball.

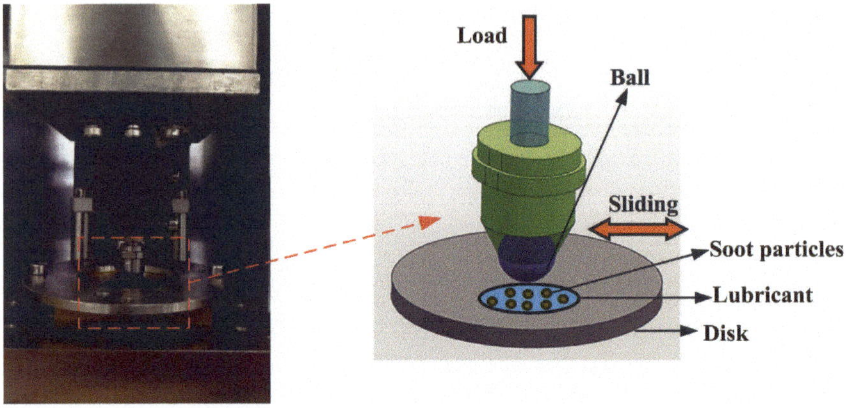

Fig. 6.1 Schematic of the friction pairs of ball-on-disk reciprocating tribometer.

A thermogravimetry analysis (TGA) of the BDS was carried out using a thermal analyzer (STA449F3, Netzsch) from room temperature up to 700°C at a ramping rate of 20°C/min in an air atmosphere. The particle size distributions of the BDS and T-BDS-OLA in LP were determined using a zeta potentiometer (Nano-ZS90, Malven). BDS, T-BDS, and T-BDS-OLA were characterized using a Raman spectroscope (HR Evolution, Horiba Jobin Yvon), field-emission transmission electron microscope (FETEM, JEM-2100F, JEOL), and Fourier transform infrared spectrometer (FTIR, Nicolet6700, Thermo Nicolet) at 500–4000/cm. The wear zones on the upper ball samples were analyzed using a 3D laser-scanning microscope (VK-X100, Keyence), scanning electron microscope coupled with an energy dispersive spectroscope (SEM/EDS, SU8010, Hitachi), and Raman spectroscopy.

6.1.3 Characterization

Fig. 6.2 shows the TG curve of the BDS, where two weight-loss stages can be observed. A slight weight loss can first be observed at 27–150°C, which corresponds to the removal of absorbed water on the surface. Another dominant weight-loss step was observed from 450 to 675°C, caused by the oxidation of BDS. The surface oxidation of the BDS occurred at a temperature range of 450–500°C. A surface oxidation treatment is beneficial to deriving a significant number of oxygen-containing functional groups on the BDS surface and changing the degree of order of the BDS [16, 17]. Moreover, a deep oxidation of the BDS occurred at a temperature of 500–675°C. Therefore, the thermal oxidation treatment temperature was chosen to be 500°C.

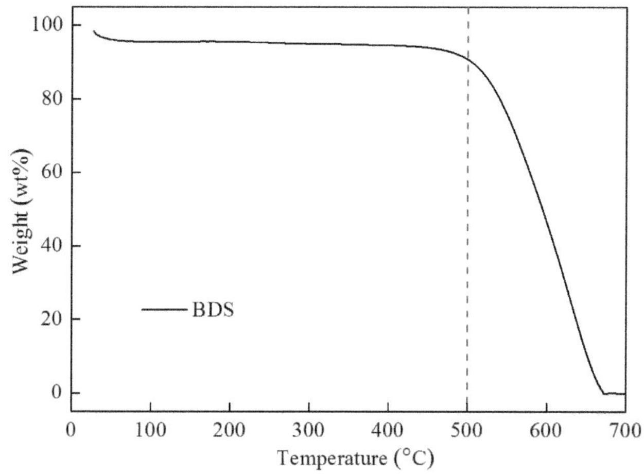

Fig. 6.2 TG curve of BDS.

Fig. 6.3 shows FETEM images of BDS, T-BDS, and T-BDS-OLA. The primary particle size of the BDS, T-BDS, and T-BDS-OLA was ~35 nm. Fig. 6.3A indicates that the BDS contains some short, disconnected, and concentrically oriented graphitic segments, which are surrounded by amorphous phases. Fig. 6.3B and C show that T-BDS and T-BDS-OLA contain more graphitic segments, which are highly curved, discontinuous, and similar to the onion-ring structure. Compared with BDS, T-BDS and T-BDS-OLA exhibit higher-ordered carbon nanostructures. The results indicate that the oxidation treatment at 500°C can improve the degree of the order of the BDS [10].

Fig. 6.4 shows the Raman spectra of BDS, T-BDS, and T-BDS-OLA. A D (disordered graphitic lattice) band and G (graphitic lattice) band appear at ~1345/cm and ~1590/cm, respectively. The I_D/I_G ratio reflects the degree of graphitization disorder of the carbon materials [18], with decreases in the ratio correlating with decreases in the amount of disorder. The results show

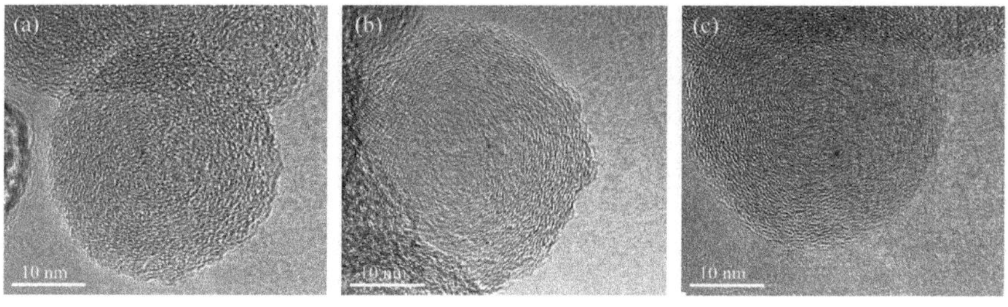

Fig. 6.3 FETEM images: (A) BDS, (B) T-BDS, and (C) T-BDS-OLA.

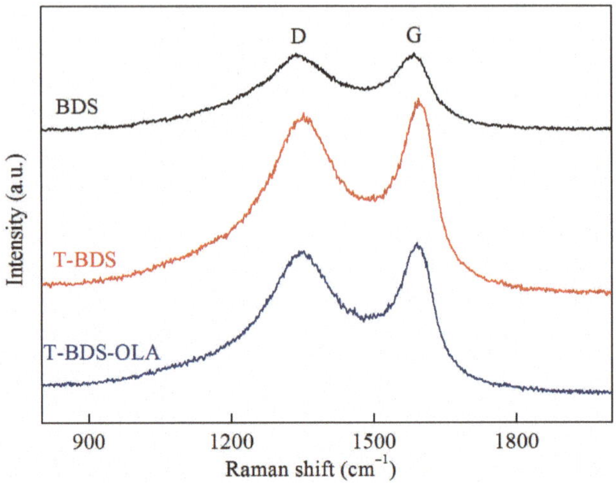

Fig. 6.4 Raman spectra of BDS, T-BDS, and T-BDS-OLA.

that the I_D/I_G ratio of T-BDS (2.639) is similar to that of T-BDS-OLA (2.656). However, the I_D/I_G ratio of T-BDS and T-BDS-OLA is lower than BDS (2.937). This means that T-BDS and T-BDS-OLA have a lesser degree of graphitization disorder than BDS. These results were confirmed through FETEM images of the soot samples, shown in Fig. 6.3.

Fig. 6.5 shows the FTIR spectra of oleylamine (OLA), BDS, T-BDS, and T-BDS-OLA. For the BDS, the main features in the spectrum are related to O–H vibrations (3440/cm), C=C vibrations (1600/cm), and C–O–C

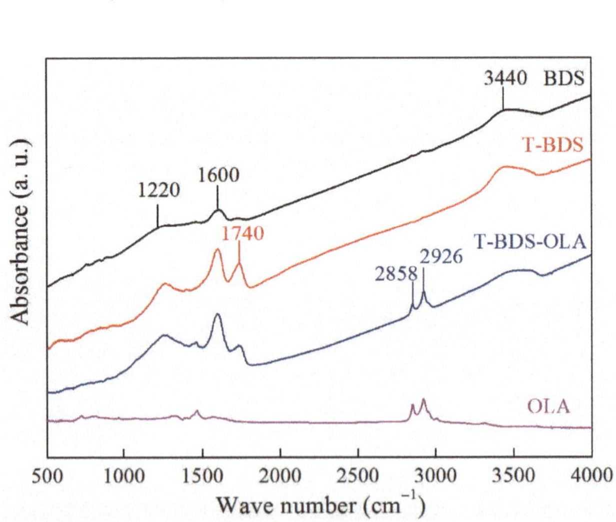

Fig. 6.5 FTIR spectra of OLA, BDS, T-BDS, and T-BDS-OLA.

stretching (1220/cm) [19]. T-BDS showed almost the same FTIR absorption peaks as BDS, but with an apparent peak of C=O (1740/cm) vibrations. The apparent peak of C=O vibrations indicates that T-BDS has been oxidized. Further, the FTIR absorption peaks of T-BDS-OLA show features of OLA at 2858 and 2926/cm ($-CH_3$ and $-CH_2-$ stretch) [9], indicating that OLA was effectively adsorbed on the surface of the soot particles and that the surface of T-BDS-OLA contained long carbon chain lipophilic groups.

Fig. 6.6 shows the particle size distribution of BDS and T-BDS-OLA in LP based on a zeta potentiometer analysis. The results show that the average particle size of BDS in LP was 203.1 nm within a range of 122.4–342 nm,

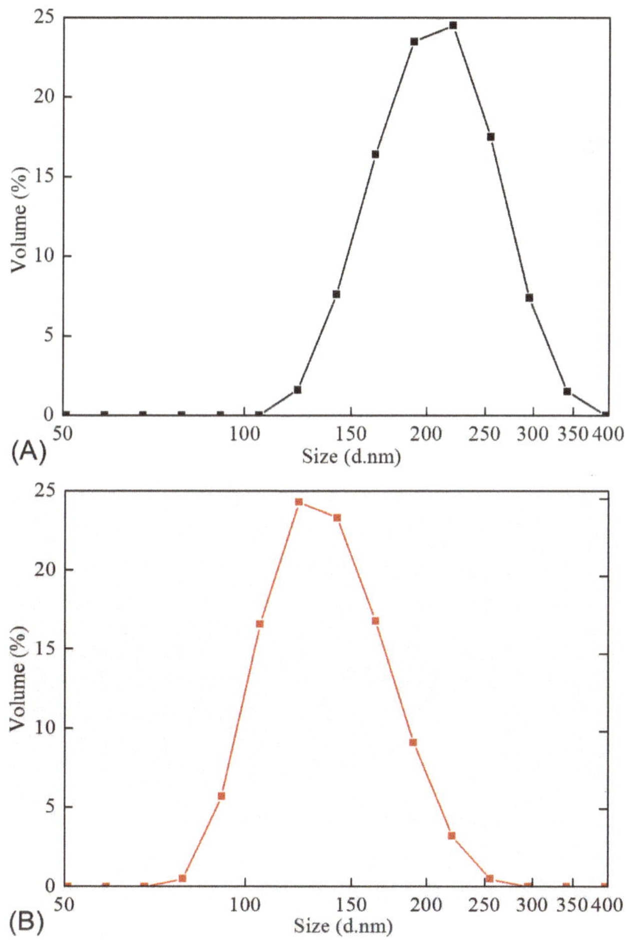

Fig. 6.6 Particle size distribution of (A) BDS in LP and (B) T-BDS-OLA in LP.

and that the average particle size of T-BDS-OLA in LP was 146.2 nm within a range of 78.82–255 nm. This means that the average particle size of T-BDS-OLA in LP was smaller than that of BDS, which was due to the fact that the lipophilic group of T-BDS-OLA impeded the particle aggregation, and stabilized it in the LP [20].

6.1.4 Tribological behavior

Fig. 6.7 shows the variations in mean friction coefficient and mean wear volumes lubricated with different amounts of the two types of soot in LP, respectively, at a load of 100 N, speed of 50 mm/s, and sliding time of 30 min. As shown in Fig. 6.7A, with an increase in content of the two types of soot, the mean friction coefficient of the soot in LP showed a decreasing trend. The mean friction coefficient of LP was 0.205. By contrast, the mean friction coefficients of LP with 0.3 wt% BDS and LP with 0.3 wt% T-BDS-OLA were 0.154 and 0.149, respectively. When the amount of the two types of soot exceeded 0.3 wt%, their average friction coefficient in LP tended to be stable. The mean wear volumes of the two types of soot in LP were less than that of LP itself, as shown in Fig. 6.7B. The mean wear volumes were 1.4×10^6, 0.72×10^6, and 0.65×10^6 μm^3 for the pure LP and LP with 0.2 wt% BDS and T-BDS-OLA, respectively. Moreover, LP with 0.3 wt% T-BDS-OLA had the lowest mean wear volumes (0.5×10^6 μm^3) in LP with different BDS or T-BDS-OLA concentrations. However, when the amount of the two types of soot was over a critical level (0.2 wt% for BDS, 0.3 wt% for T-BDS-OLA), the mean wear volumes of the two types of soot in LP showed a slightly increasing trend. This demonstrates that BDS and T-BDS-OLA favor a reduction in the friction and an improvement in the wear resistance of LP when the amount of the two types of soot is low (0.1 wt% to 0.4 wt%) [21]. However, an excessive amount of both types of soot is not conducive to improving the wear resistance of the soot in LP [3]. In addition, under the same amount, the average friction coefficient and mean wear volumes of T-BDS-OLA in LP were lower than those of BDS. The results confirmed that T-BDS-OLA in LP possesses a better friction reduction and wear resistance than BDS in LP at different amounts.

Fig. 6.8 shows the variations in mean friction coefficient and mean wear volumes lubricated with 0.3 wt% of the two types of soot in LP at different loads, a speed of 50 mm/s, and a sliding time of 30 min. As shown in Fig. 6.8A, when the load is increased, the average friction coefficient of LP first exhibits a slight decrease and then increases. When the load exceeds 100 N, the average friction coefficient of LP remains mostly stable.

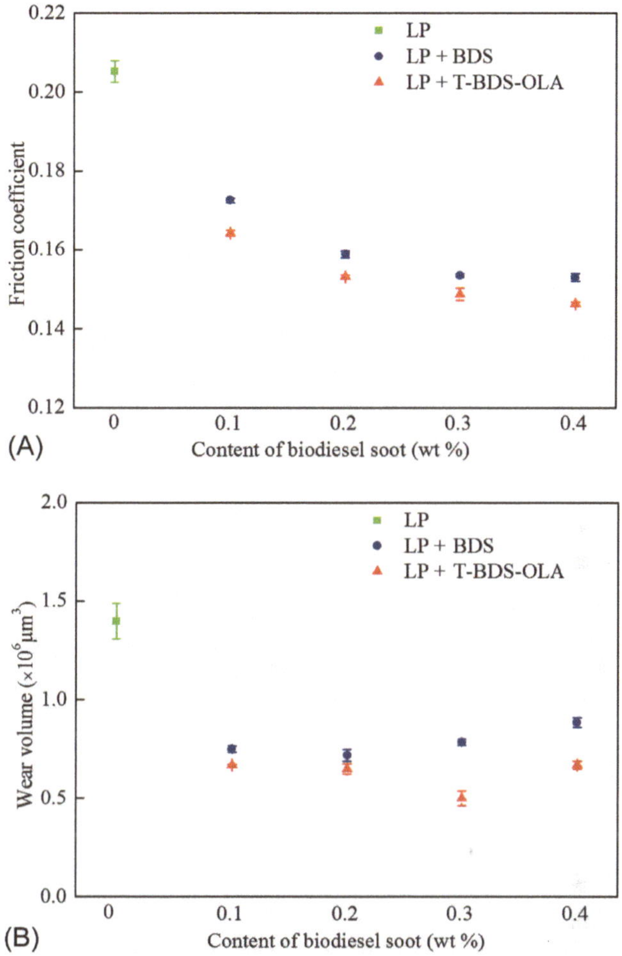

Fig. 6.7 Variations of (A) mean friction coefficient and (B) mean wear volumes when lubricated with different amounts of the two types of soot in LP at 100 N and 50 mm/s for 30 min.

Moreover, the average friction coefficients of the two types of soot in LP are less than that of the LP itself, particularly at a high load (100 or 150 N). The mean wear volumes of the two types of soot in LP are also remarkably decreased in comparison with the LP itself, as indicated in Fig. 6.8B. This may be due to the formation of a lubricant film by LP under a low load. However, with an increase in load, the lubricant film is easy to destroy, resulting in an enhancement of the direct joggle effect between the microconvex bodies on the friction surface and the aggravation of friction

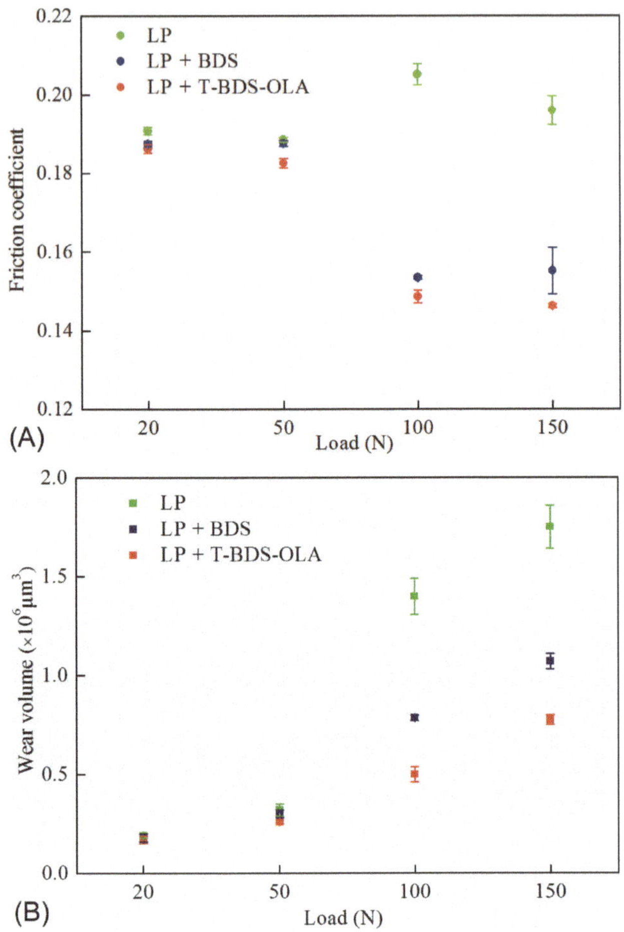

Fig. 6.8 Variations of (A) mean friction coefficient and (B) mean wear volumes lubricated with 0.3 wt% of the two types of soot in LP at different loads and 50 mm/s for 30 min.

and wear [9]. The two types of soot can improve both the antifriction and antiwear properties of LP, and were observed to have better antifriction and antiwear properties at a higher load (100 or 150 N) compared with at a lower load (20 or 50 N). In addition, when the BDS was replaced with T-BDS-OLA, the average friction coefficient and the mean wear volumes of the LP were further improved. When the load was 100 N, the mean friction coefficients of LP with BDS and T-BDS-OLA were reduced by 24.9% and 27.3%, respectively, compared to LP. As for the addition of BDS and T-BDS-OLA, the mean wear volumes of LP were decreased by 43.9% and

64.3%, respectively, with an applied load of 100 N. This finding shows that T-BDS-OLA with LP possesses better antifriction and antiwear properties than BDS with LP under different loads.

6.1.5 Analysis of worn surface

Fig. 6.9 shows the characterization results of the wear zones for LP, LP containing 0.3 wt% BDS, and LP containing 0.3 wt% T-BDS-OLA, at a load of 100 N, speed of 50 mm/s, and sliding time of 30 min. As shown in the FESEM image in Fig. 6.9A, many large spalling pits occurred in the wear zones lubricated with LP, and as shown in Fig. 6.9B, many furrows and small pits can be observed in the wear zones lubricated with BDS in LP. In addition, the FESEM image of the wear scars lubricated with T-BDS-OLA in LP provides no information regarding the pits, and only slight furrows can be seen in Fig. 6.9C. These results are confirmed through 3D microscopic images and profiles of the wear zones. With LP, the profile curve was both the deepest and the broadest (Fig. 6.9D). With BDS in LP, the profile curve became narrow and shallow (Fig. 6.9E). Fig. 6.9F shows that the profile curve of T-BDS-OLA in LP became narrower and shallower than BDS in LP.

This indicates that severe spalling wear occurred for the wear zones lubricated with LP. BDS can improve the bearing capacity of an LP film and can fill in the friction surface to avoid direct contact between the micro-convex bodies [22], reducing the spalling wear. Meanwhile, BDS can also roll on the friction surface. Nevertheless, the aggregation of BDS aggravates abrasive wear [23]. The lipophilic group of T-BDS-OLA inhibits such aggregation, thereby reducing the abrasive wear. These characterizations are described in Figs. 6.5 and 6.6. Apart from spacing and rolling on the surface, T-BDS-OLA has a carbon onion ring-like structure, which is prone to exfoliation into individual graphitic sheets, producing an extraordinary lubrication effect [24]. These results are consistent with the trends shown in Figs. 6.7 and 6.8.

6.1.6 Lubrication mechanism

To further study the tribological differences between BDS and T-BDS-OLA, a Raman analysis was carried out on the wear zones. Fig. 6.10 shows the Raman analysis results of the wear zones of the upper balls after the tribo-test. As Fig. 6.10 indicates, for the two types of soot in LP used in our study, two typical characteristic peaks of soot at the D and G bands can clearly be seen. This observation suggests that the two types of soot were

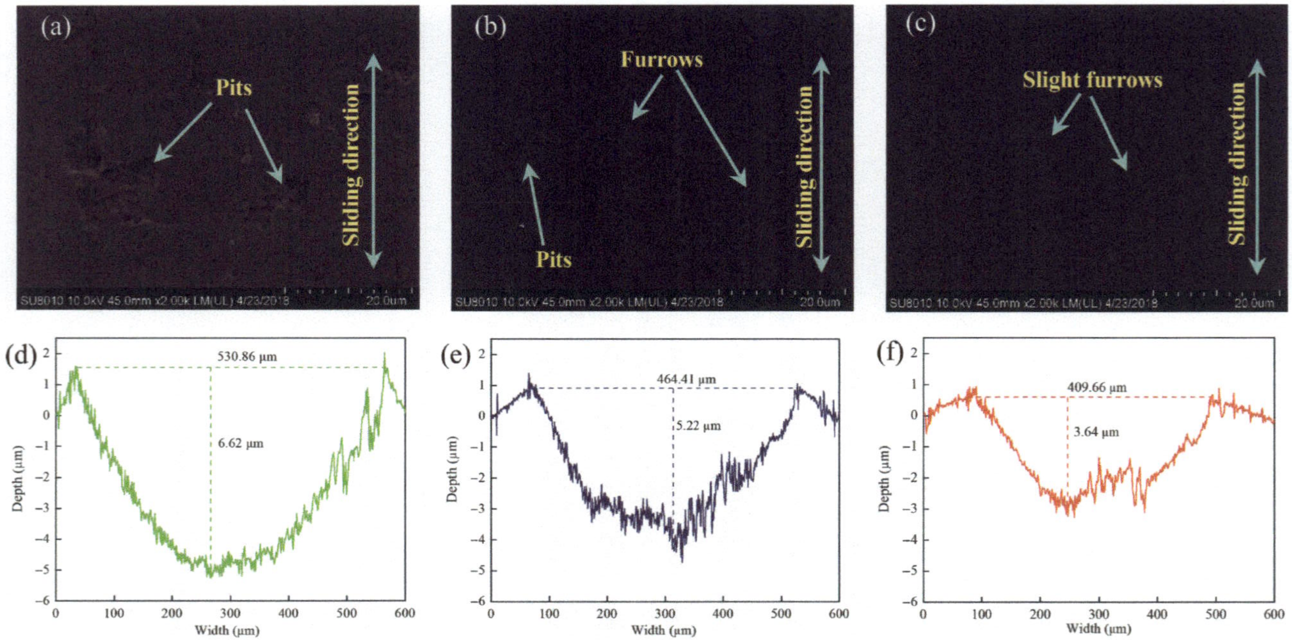

Fig. 6.9 FESEM images and profiles of the wear zones of the upper balls for (A and D) LP, (B and E) LP containing 0.3 wt% BDS, and (C and F) LP containing 0.3 wt% T-BDS-OLA at 100 N and 50 mm/s for 30 min.

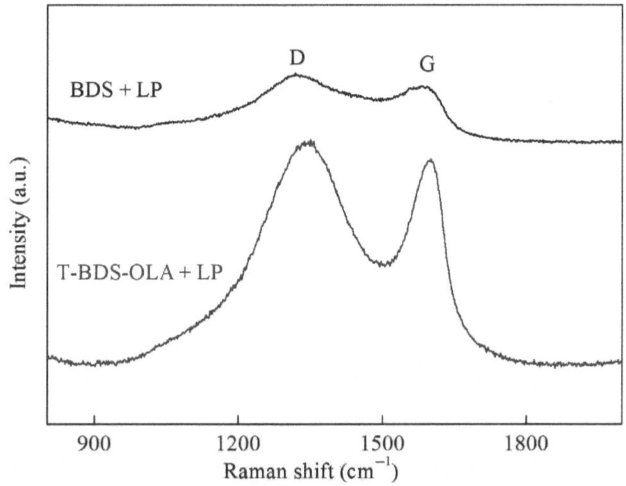

Fig. 6.10 Raman analysis of wear zones of upper balls after tribo-test.

adsorbed on the surface of the steel substrate to prevent direct contact of the friction pairs [25]. Combined with the results in Fig. 6.4, the I_D/I_G ratio of the wear zone lubricated with BDS in LP (2.927) is shown to be similar to the I_D/I_G ratio of BDS (2.937). However, the I_D/I_G ratio of the wear zone lubricated with T-BDS-OLA in LP (2.900) was larger than that of T-BDS-OLA (2.656). This indicates that the degree of order of BDS showed little change after the tribo-test. However, the degree of order of T-BDS-OLA showed a larger change. This can be explained through the exfoliation of graphitic sheets from T-BDS-OLA, which is responsible for the better lubricating performance compared with BDS [26].

Based on the observations described earlier, the lubrication mechanisms of BDS and T-BDS-OLA in LP are considered to be as follows. First, both types of soot can fill in the friction interfaces to avoid direct contact between the microconvex bodies during the tribological test [27]. Second, both types of soot can roll on the friction surface, thereby improving the friction-reducing properties. However, BDS aggregates more easily than T-BDS-OLA, and the aggregation of soot particles causes an increase in abrasive wear [28, 29]. Furthermore, under a high load and shear action, some of the graphitic sheets are exfoliated from T-BDS-OLA, forming a carbon lubricating layer between the rubbing surfaces, which slides easily and prevents contact between the surfaces, thereby helping reduce the friction and wear of the friction surfaces [25, 30].

6.1.7 Conclusions

In summary, thermally oxidized treatment at 500°C can increase the order of BDS, and a modification using oleylamine can thereby improve its oil solubility. FETEM and Raman spectra revealed that the degree of order of T-BDS-OLA was better than that of BDS. The graphitic layers in T-BDS-OLA are similar to a carbon onion-ring structure. The C=O vibrations and $-CH_3$ and $-CH_2-$ stretching in the FTIR spectra suggest that T-BDS is oxidized and that the surface of T-BDS-OLA contains long carbon chain lipophilic groups, respectively. The average particle size of T-BDS-OLA (146.2 nm) in LP is smaller than that of BDS (203.1 nm).

BDS and T-BDS-OLA can both improve the antifriction and antiwear properties of LP when used in low amounts (0.1–0.4 wt%). Moreover, T-BDS-OLA with LP has better antifriction and antiwear properties than BDS with LP. The tribological mechanisms can be summarized as follows: (1) During a tribological test, BDS and T-BDS-OLA both enter the friction interfaces and prevent direct contact between the friction pairs; (2) The two types of soot can act as ball bearings between the contact areas; (3) Exfoliated graphitic sheets from T-BDS-OLA can form a carbon-lubricating layer between the friction surfaces that slides easily and thus avoids contact between the friction surfaces. Thus, T-BDS-OLA in LP presents better tribological properties.

6.2 Novel carbon nanoparticles derived from biodiesel soot as lubricant additives

6.2.1 Introduction

With changes in the global climate and growing energy demands, there has been an increased interest in renewable fuels to replace petroleum. Biodiesel is a potential renewable fuel source. Biodiesel (fatty acid alkyl ester) can be obtained via the transesterification process [10]. It exhibits characteristics comparable to those of petroleum diesel, as well as low sulfur, low aromatics, low viscosity, high flash point, high lubricity, and high cetane number [31]. Although biodiesel has a higher oxygen content than petroleum diesel, which enhances its combustion efficiency, biodiesel soot (BDS) is still emitted [32, 33].

Soot can contaminate the lubricating oil within the sump as a result of the blow-by gases [34]. Soot in the lubricating oil can increase the viscosity of the oil and affect its tribological properties. Salehi et al. [15] found that rapid tribofilm formation through the addition and removal of ZDDP (zinc dialkyl dithiophosphates) by soot particles results in severe wear, which can be attributed to the corrosive–abrasive mechanism. Green et al. [35] concluded that a larger soot amount in the lubricating oil generates more wear, which is mainly due to the abrasive process. However, soot also has potential as a lubricant additive. Guo et al. [5] reported that the antiwear and friction-reduction properties of PAO (polyalphaolefin) 4 oil were improved by adding 0.01 wt% diesel soot. Additionally, carbon nano-onions have been widely studied as lubrication additives because they can improve the tribological properties of oils. Joly-Pottuz et al. [36] found that carbon nano-onions have no dangling bonds on their surface and can therefore easily slide and even roll on a friction surface. Hirata et al. [12] reported that the closed structure of carbon nano-onions provides good mechanical strength. BDS contains short graphene segments of high tortuosity [37, 38], which exhibit excellent potential for the development of onion-like carbon nanoparticles. However, to the best of our knowledge, few reports have described onion-like carbon nanoparticles derived from BDS as a lubricant additive.

At present, real engine soot is expensive and not widely available because it can only be produced by running an engine for a long time under suboptimal conditions [39]. Thus, engine soot alternatives (such as carbon black and self-prepared soot) are typically used to research the effect of engine soot on lubricating oils [40, 41]. Clague et al. reported that diesel engine soot and carbon black exhibited similar primary particle sizes and internal structures of particulates [42]. Ferraro et al. used various techniques to compare carbon black and diesel engine soot and found that carbon black was an appropriate substitute for engine diesel soot [43]. Alternatively, self-prepared soot can be used as an engine-soot alternative. Self-prepared soot was prepared via the combustion of fuel at a normal temperature and atmospheric pressure. Hu et al. compared the differences between carbon black and self-prepared biomass-oil soot. Differences were observed in the elemental composition and the functional groups. However, self-prepared biomass-oil soot is similar in size, morphology, and internal structure to carbon black. Studies have indicated that self-prepared BDS has discontinuous graphitic layers that are surrounded by an amorphous phase. This structure is similar to that of real BDS from a diesel engine [44, 45]. Thus, self-prepared BDS can be used as an alternative to real BDS.

In this study, onion-like carbon nanoparticles derived from BDS were prepared. The tribological properties of nitric acid-treated BDS (NA-BDS) as an additive in water (H_2O) and NA-BDS modified with oleylamine (NA-BDS-OLA) as an additive in liquid paraffin (LP) were investigated.

The corresponding tribological mechanisms were also examined. Thus the objective of this study was to provide a new path for recycling carbon resources from engine emissions and for reducing environmental pollution from fuel combustion.

6.2.2 Experiments

6.2.2.1 Sample preparation

All reagents were analytical grade, including methanol (Sinopharm Chemical Reagent Co., Ltd., Shanghai, China), oleylamine (Aladdin Industrial Co., Ltd., Shanghai, China), ethanol (China Sun Specialty Products Co., Ltd., Changshu, China), and nitric acid (68%, Sinopharm Chemical Reagent Co., Ltd., Shanghai, China). Soybean oil was obtained from China Oil and Foodstuffs Corporation (Beijing, China). Ultrapure water obtained from a water purification system (UPT-11-10T, Sichuan Ulupure Super Pure Technology Co., Ltd., Chengdu, China) was used in all experiments. Biodiesel was synthesized through transesterification between methanol and soybean oil using a calcium-based catalyst. The physical properties are presented in Table 1. The procedure for collecting BDS was as follows [46, 47]: (1) First, 80 cm^3 of biodiesel was poured into a 100 cm^3 ceramic crucible, and then the ceramic crucible was placed on an iron stand. (2) Clean glass slides (BDS capturing slides) were used as substrates to collect the BDS. The glass slides were clamped on the upper part of the iron stand, above the ceramic crucible. When the biodiesel was ignited, the slide was kept over the flame at a distance of 2 cm. (3) As the biodiesel burned, the BDS in the flame was deposited onto the slide surface. When the biodiesel in the ceramic crucible was completely burned and a BDS layer with a thickness of a few micrometers was deposited on the glass slide, the BDS layer was carefully removed using a laboratory spoon and collected in a glass bottle. (4) Finally, the collected BDS was dried at 120°C for 3 h and ground into a fine granular powder. Burning 1 kg of biodiesel yielded 13–18 g of BDS.

A total of 0.2 g of BDS was suspended in 30 g of concentrated nitric acid (68%, analytical reagent) and refluxed for 20 h in an oil bath maintained at 80°C. After repeated centrifugation and washing with ultrapure water, it was then dried at 120°C for 10 h. This product was denoted as NA-BDS. Then, 0.1 g of the NA-BDS was refluxed in 5 g of oleylamine (OLA) at 110°C for 12 h. The mixture was purified through repeated washings with ethanol using centrifugation. Finally, it was dried at 120°C for 5 h. This product was denoted as NA-BDS-OLA. As additives, BDS and NA-BDS (at concentration levels of 0.1, 0.2, and 0.3 wt%) were added to the H_2O.

Moreover, BDS and NA-BDS-OLA (at concentration levels of 0.2, 0.3, and 0.4 wt%) were added to the LP. The mixtures were then distributed using a glass rod for 15 min, followed by magnetic stirring for 1 h to reduce any deviations in the experiment.

6.2.2.2 Tribological testing and characterization

Tribological testing was performed using a ball-on-disc reciprocating tribometer (CFT-I, Lanzhou Zhongke Kaihua Technology Development Co. Ltd, Lanzhou, China). The upper ball samples (Φ 6 mm, surface roughness of $R_{a1} = 0.02$ μm) and lower disk samples (50 × 50 × 5 mm, surface roughness of $R_{a2} = 0.5$ μm) were made of GCr15 steel with a hardness of 62–65 HRC (Rockwell hardness value). The lower disk samples were fixed, and the upper ball samples were slid in a reciprocating manner with a sliding distance of 5 mm. The friction pairs of the ball-on-disk reciprocating tribometer are illustrated in Fig. 6.1. The wear rate is estimated by Archard's equation. As the initial roughness of the lower disk samples was high, the wear volume of disc samples was very difficult to assess [15]. Therefore, the wear volume of upper ball samples was used to calculate the wear rate. In order to determine the wear rate of the upper ball samples, the following Eqs. (6.3)–(6.5) were used [48]:

$$V = \pi \times h \times \left(0.75d^2 + h^2\right) / 6, \tag{6.3}$$

$$h = R - \left(R^2 - 0.25d^2\right)^{0.5}, \tag{6.4}$$

$$K_B = V / \left(L \times F\right), \tag{6.5}$$

where K_B represents ball wear rate (in $mm^3/N \cdot m$), d represents the diameter of the wear scar of the upper ball (in mm), h represents the height of the wear of the upper ball (in mm), R represents the radius of the upper ball (in mm), V represents the wear volume of the upper ball sample (in mm^3), L represents the sliding distance (in m), and F represents the applied load (in N). Each tribological test was repeated three times to obtain a standard deviation and to reduce human error.

For the ball-on-disc contact, the Hertz contact diameter and the contact stress were calculated using Hertz's theory [49]:

$$a = 2 \times \left(\frac{2}{3} \times \frac{W \times R}{E'}\right)^{\frac{1}{3}}, \tag{6.6}$$

$$P = \frac{4 \times W}{\pi \times a^2}, \tag{6.7}$$

where a represents the Hertz contact diameter, W represents the normal load (=20, 50, and 100 N), R represents the equivalent radius of curvature (3 mm), and E' represents the effective elastic modulus (233 GPa). Thus the Hertz diameter was 1.11×10^{-4} m, 1.51×10^{-4} m, and 1.90×10^{-4} m, and the contact stress was 2, 2.8, and 3.5 GPa.

The λ was calculated using Dowson and Hamrock's minimum film thickness formula:

$$h_{min} = 3.63 \times R \times \frac{G^{*0.49} \times U^{*0.68}}{W^{*0.073}} \left(1 - e^{-0.68k}\right), \tag{6.8}$$

$$\lambda = \frac{h_{min}}{\sqrt{Ra_1^2 + Ra_2^2}}, \tag{6.9}$$

$$G^* = \alpha \times E', \tag{6.10}$$

$$U^* = \frac{\eta_0 \times U}{E' \times R}, \tag{6.11}$$

$$W^* = \frac{W}{E' \times R^2}, \tag{6.12}$$

where R represents the equivalent radius of curvature (=3 mm), k is the elliptical parameter (≈1), η_0 represents the dynamic viscosity of water or LP (1×10^{-3} or 100.2×10^{-3} Ns/m^2, respectively), U represents the sliding speed (=50 mm/s), E' represents the effective elastic modulus (=233 GPa), α is the coefficient of viscosity–pressure of water and liquid paraffin, respectively, (2.2×10^{-8} m^2/N or 2×10^{-8} m^2/N) [50, 51], and R_{a1} and R_{a2} represent the surface roughness values of the ball and disc, respectively ($R_{a1} = 0.02$ μm, $R_{a2} = 0.5$ μm).

The maximum value of λ ($W = 20$ N) was calculated to be 0.0019 and 0.0418 under water and LP lubrication, respectively, for the test conditions. This confirms that the sliding tests were conducted in the boundary lubrication regime.

The morphologies of BDS, NA-BDS, and NA-BDS-OLA were examined using Raman spectroscopy (HR Evolution, Horiba Jobin Yvon, Paris, France), field-emission transmission electron microscopy (FETEM, JEM-2100F, JEOL, Tokyo, Japan), and Fourier transform infrared (FTIR) spectroscopy (Nicolet6700, Thermo Nicolet, Madison, WI, United States). The particle-size distributions of BDS and NA-BDS in H_2O, as well as those of BDS and NA-BDS-OLA in LP, were characterized using a zeta potentiometer (Nano-ZS90, Malvern, Worcestershire, United Kingdom). After friction and wear tests, the friction surfaces were inspected and analyzed using field-emission scanning electron microscopy (FESEM, SU8010, Hitachi, Tokyo, Japan), three-dimensional laser scanning microscopy (VK-X100, Keyence, Osaka, Japan), and Raman spectroscopy.

6.2.3 Characterization

To investigate the effects of nitric acid treatment and oleylamine (OLA) modification on BDS, Raman spectroscopy, FETEM, FTIR spectroscopy, and particle-size distribution analyses were conducted. The Raman spectra of BDS, NA-BDS, and NA-BDS-OLA are presented in Fig. 6.11. The Raman spectra exhibited well-defined D and G peaks. The D peak (at ~1346/cm) is attributed to the defects and disordered graphitic lattices. The G peak (at ~1589/cm) is attributed to the graphitic lattices [52, 53]. In general, the I_D/I_G ratio increased with the increase in the degree of disorder. A comparison of the Raman spectra of BDS and NA-BDS indicated that the I_D/I_G ratio of the NA-BDS (2.546) was lower than that of BDS (2.956). The I_D/I_G ratio of NA-BDS-OLA (2.559) was similar to that of NA-BDS (2.546). The spectroscopic results indicated that nitric acid treatment improved the order of BDS.

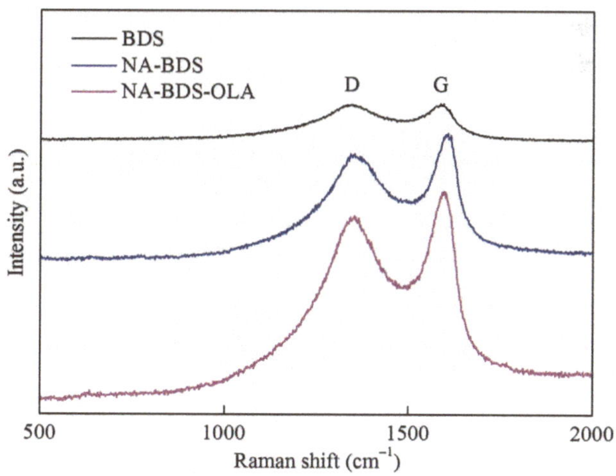

Fig. 6.11 Raman spectra of biodiesel soot (BDS), nitric-acid-treated BDS (NA-BDS), and NA-BDS modified with oleylamine (NA-BDS-OLA).

The Raman results were complemented with FETEM observations of BDS, NA-BDS, and NA-BDS-OLA, as shown in Fig. 6.12. The diameters of BDS, NA-BDS, and NA-BDS-OLA were approximately 35–40 nm. For BDS, short, disconnected, and concentrically oriented graphitic layers were observed, as shown in Fig. 6.12A. However, these graphitic layers were surrounded by amorphous phases. As shown in Fig. 6.12B and C, compared with BDS, the apparent growth of highly curved and discontinuous graphitic layers was observed in NA-BDS and NA-BDS-OLA, respectively, which were structurally similar to onion-like carbon. A statistical evaluation of the FETEM micrographs indicated that the degree of graphitization was higher for NA-BDS and NA-BDS-OLA than for BDS, in agreement with the Raman results [10].

As shown in Fig. 6.13, the functional groups on the surfaces of OLA, BDS, NA-BDS, and NA-BDS-OLA were investigated using FTIR spectroscopy. For

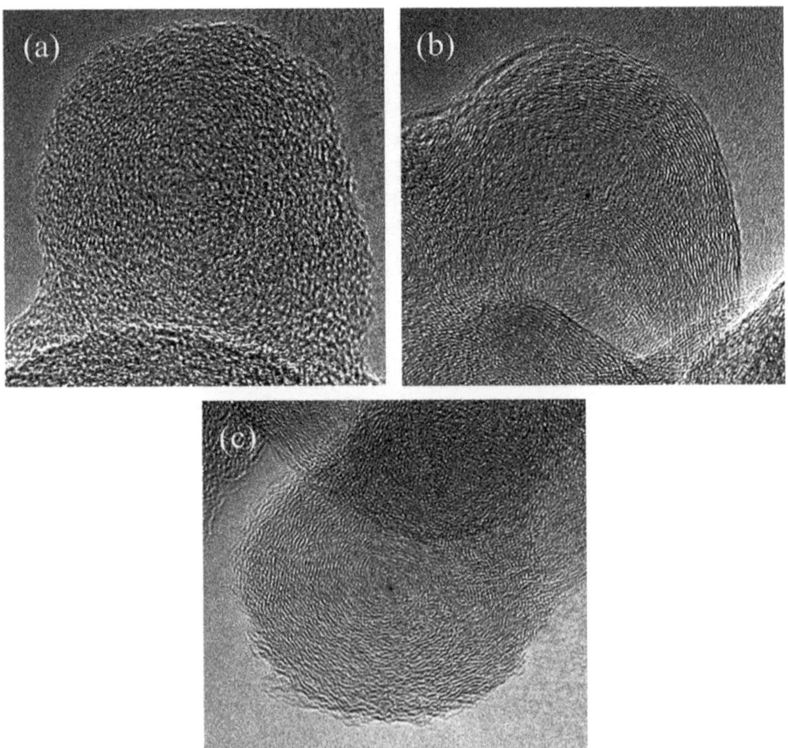

Fig. 6.12 Field-emission transmission electron microscopy (FETEM) micrographs of (A) BDS, (B) NA-BDS, and (C) NA-BDS-OLA

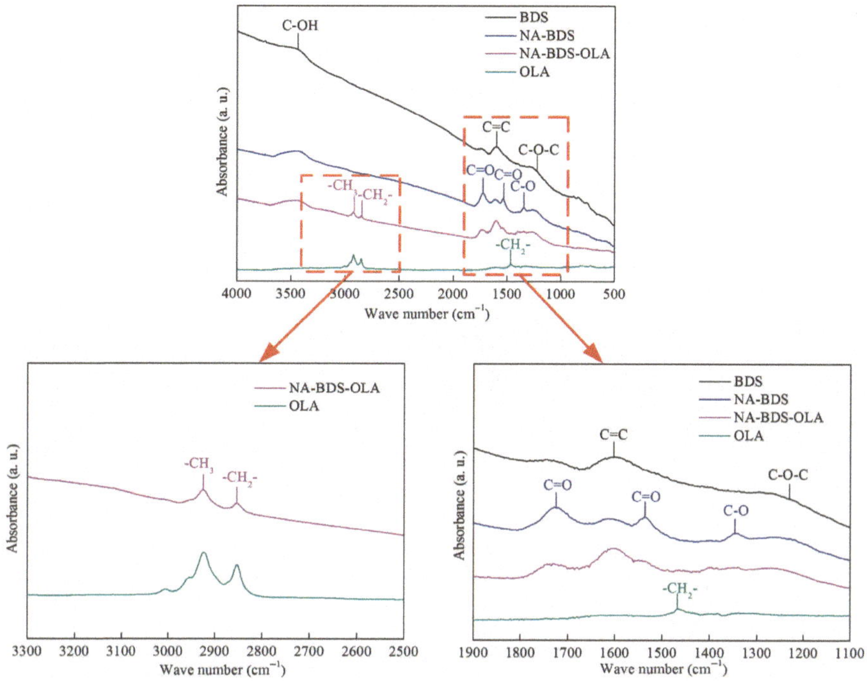

Fig. 6.13 Fourier transform infrared (FTIR) spectra of oleylamine (OLA), BDS, NA-BDS, and NA-BDS-OLA.

BDS, the three main peaks at 1221, 1602, and 3443/cm are ascribed to the C–O–C, C=C, and C–OH stretching vibrations, respectively. For NA-BDS, the peaks at (1535 and 1726/cm) and 1347/cm are attributed to the stretching vibrations of C=O and C–O, respectively, indicating that strong (–COOH) polar groups were introduced into the NA-BDS through the nitric acid treatment [10]. For OLA, the main peaks in the spectrum are related to $-CH_2-$ (1465 and 2853/cm) and $-CH_3$ (2923/cm). Moreover, for NA-BDS-OLA, the peaks at 2853 and 2923/cm are associated with $-CH_2-$ and $-CH_3$ stretching vibrations, respectively, indicating that the surface of NA-BDS-OLA contained long-carbon-chain lipophilic groups owing to the OLA treatment [54]. This was because of the many active hydroxyl and carboxyl groups of NA-BDS, which can adsorb and react with OLA. Interestingly, the peak shape of C–H under NA-BDS-OLA became narrower than OLA, indicating that the C–H content of NA-BDS-OLA was smaller than OLA.

The particle-size distributions of the different soot samples are presented in Fig. 6.14. As shown in Fig. 6.14A, the particle size of BDS in H_2O ranged from 91.3 to 615.1 nm, with a mean value of 219.7 nm.

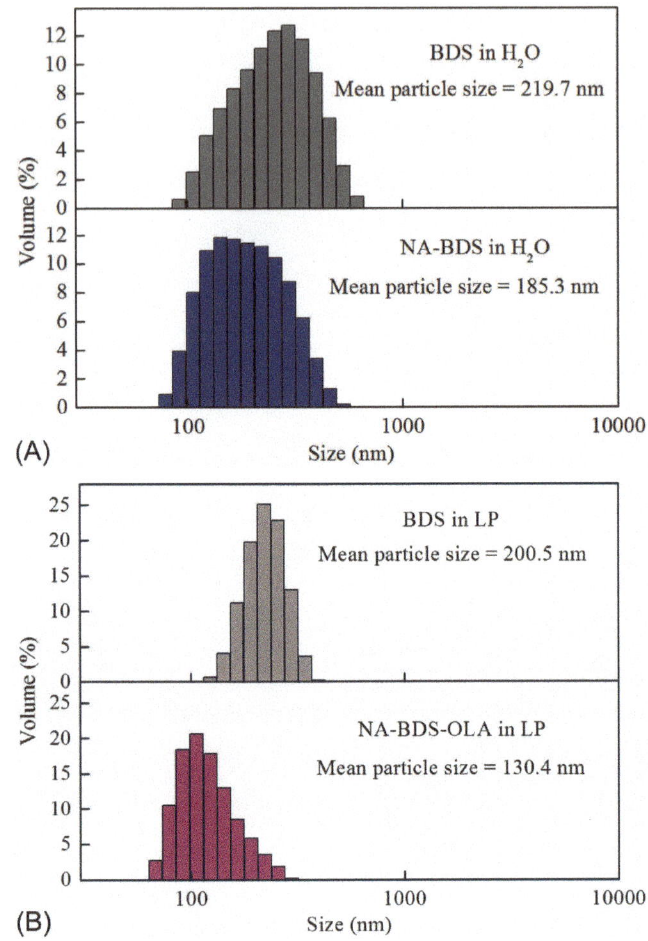

Fig. 6.14 Particle-size distributions of (A) BDS and NA-BDS in H_2O, and (B) BDS and NA-BDS-OLA in liquid paraffin (LP).

Compared with BDS, the mean particle size of NA-BDS in H_2O was reduced to 185.3 nm, whereas the particle size of NA-BDS in H_2O ranged from 78.8 to 531.2 nm. This phenomenon was due to the oxidation of BDS, which increased the amount of hydrophilic functional groups in NA-BDS [10], and stabilized the material in H_2O. As indicated by Fig. 6.14B, the mean particle size decreased from 200.5 nm (between 122.4 and 396.1 nm for BDS in LP) to 130.4 nm (between 68.1 and 295.3 nm for NA-BDS-OLA in LP), which is ascribed to the introduction of functional groups with high lipophilicity on the surface of NA-BDS-OLA [55]. These results agree well with the FTIR measurements.

6.2.4 Tribological performance of soot samples in H_2O

The evolutions of the average friction coefficient and wear rates for lubrication with different amounts of two types of soot in H_2O are plotted in Fig. 6.15. Clearly, both BDS and NA-BDS reduced the friction and wear for all the tested concentrations. H_2O + NA-BDS exhibited the best lubrication performance. At the optimum concentration of 0.2 wt%, the friction coefficients and wear rates of H_2O + NA-BDS were reduced by 52.5% and 66.9%, respectively, compared with that lubricated using H_2O. This indicates that, at different amounts, BDS and NA-BDS can improve

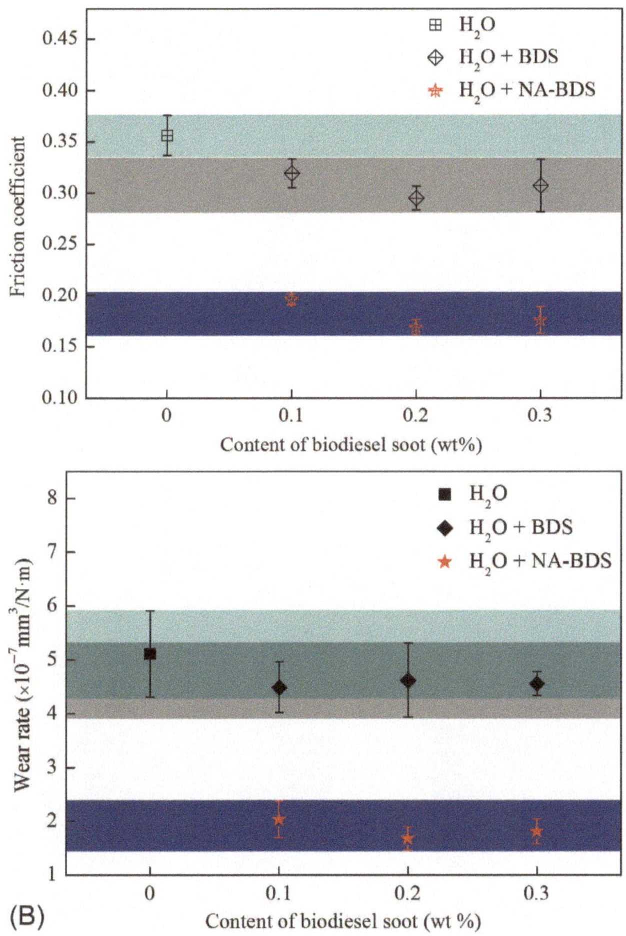

Fig. 6.15 Evolutions of the (A) average friction coefficient and (B) wear rates for lubrication with different amounts of BDS and NA-BDS in H_2O (load = 50 N, speed = 50 mm/s, duration = 30 min).

the friction and wear behaviors of H_2O, and that NA-BDS is more effective than BDS.

The evolutions of the average friction coefficient and the wear rates for lubrication with the two types of soot in H_2O at different loads are shown in Fig. 6.16. Under the different loads, the friction coefficients and wear rates decreased in the following order: $H_2O > H_2O + BDS > H_2O + NA\text{-}BDS$. Thus the antifriction and antiwear properties of H_2O + NA-BDS were better than those of H_2O + BDS and H_2O. Furthermore, BDS and NA-BDS

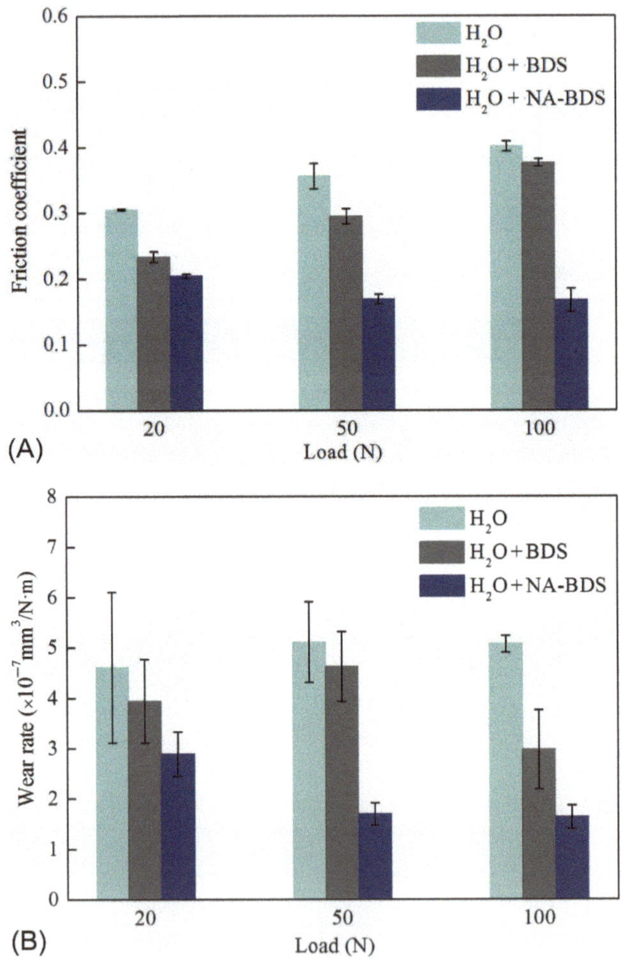

Fig. 6.16 Evolutions of the (A) average friction coefficient and (B) wear rates for lubrication using BDS and NA-BDS in H_2O at different loads (soot content = 0.2 wt%, speed = 50 mm/s, duration = 30 min).

enhanced the tribological performance of H_2O from a low to high load, indicating that carbon nanoparticles (BDS and NA-BDS) entered the rubbing surfaces and performed their function, even under a high load.

Fig. 6.17 shows FESEM images, optical micrographs, and corresponding worn surface profiles of the wear scars on the upper balls lubricated using H_2O, BDS + H_2O, and NA-BDS + H_2O. Comparing Fig. 6.17A–C reveals that the surface wear scars in the case of H_2O had

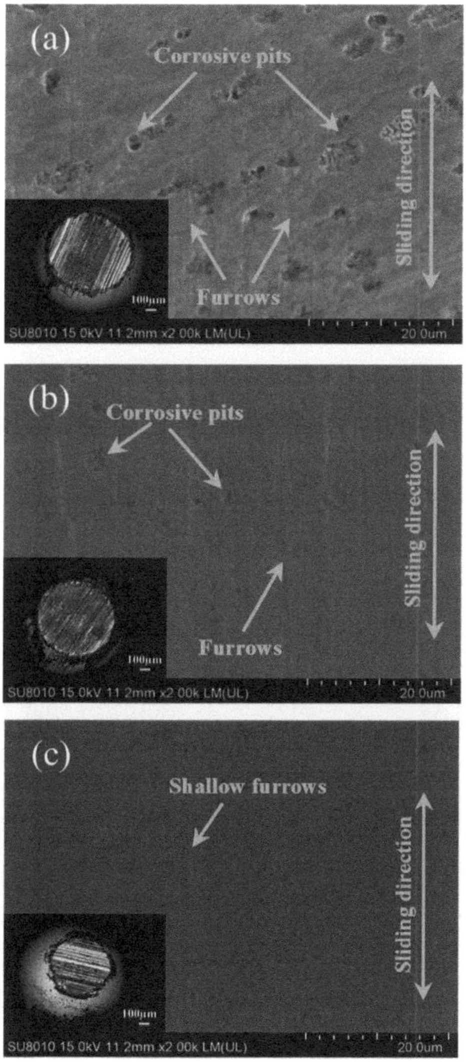

Fig. 6.17 See figure legend on next page.

(Continued)

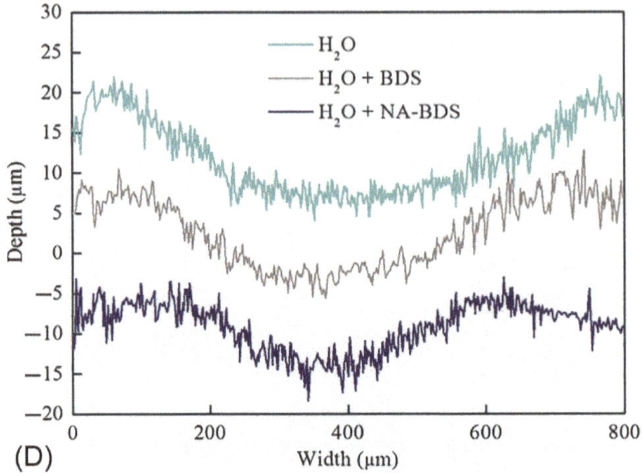

(D)

Fig. 6.17 Field-emission scanning electron microscopy (FESEM) images and optical micrographs of wear scars on upper balls lubricated using (A) H_2O, (B) H_2O + BDS, and (C) H_2O + NA-BDS. (D) Corresponding worn surface profiles (soot content = 0.2 wt%, load = 100 N, speed = 50 mm/s, duration = 30 min).

many corrosive pits and furrows [56]. When 0.2 wt% BDS was added to H_2O, although the surface wear scars still exhibited many furrows, the number of corrosive pits was reduced. The worn surfaces of the upper balls lubricated using NA-BDS + H_2O exhibited thin and shallow furrows and no corrosive pits. Compared with H_2O and H_2O + BDS, when H_2O + NA-BDS was used as the lubricant, the surface roughness *Ra* of the worn surface of the ball was lower. Moreover, Fig. 6.17D, which combines the optical micrographs in Fig. 6.17A–C, indicates that the width and depth of the wear scar decreased in the following order: H_2O > H_2O + BDS > H_2O + NA-BDS. This indicates that H_2O + NA-BDS had a better wear-resistance effect than H_2O + BDS, in agreement with the results in Fig. 6.16. Section 6.3 provides a more detailed discussion of the lubricity and lubrication mechanism of the modified biodiesel soot in water.

6.2.5 Tribological performance of soot samples in LP

The evolutions of the average friction coefficient and the wear rates for lubrication using different amounts of the two types of soot added to LP are presented in Fig. 6.18. The friction coefficients and wear rates for LP + NA-BDS-OLA and LP + BDS were lower than those for LP. LP + NA-BDS-OLA exhibited the best tribological properties. At the optimum concentration

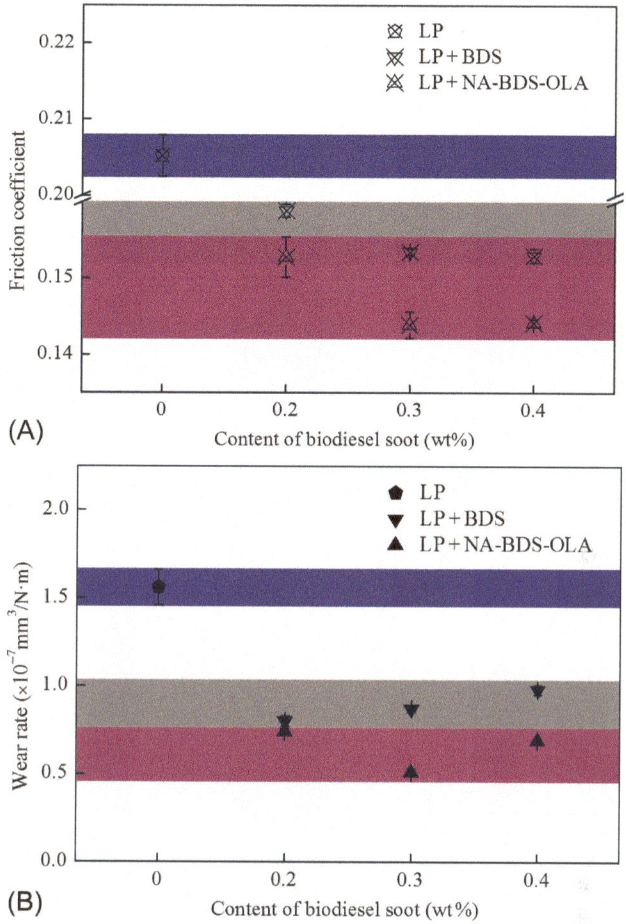

Fig. 6.18 Evolutions of the (A) average friction coefficient and (B) wear rates for lubrication with different amounts of BDS and NA-BDS-OLA in LP (load = 100 N, speed = 50 mm/s, duration = 30 min).

of 0.3 wt%, the friction coefficients and wear rates of LP + NA-BDS-OLA were 29.8% and 67.3% lower than those of LP, respectively. This indicates that at different amounts, BDS and NA-BDS-OLA can be used as lubrication additives for LP, and that LP + NA-BDS-OLA has better antifriction and antiwear properties than LP + BDS.

Fig. 6.19 shows the evolution of the average friction coefficient and wear rates for lubrication using the two types of soot in LP at different loads. The results indicate that the friction coefficients and wear rates for LP were the highest, followed by LP + BDS, and then LP + NA-BDS-OLA. Thus LP + NA-BDS-OLA exhibited the best tribological performance.

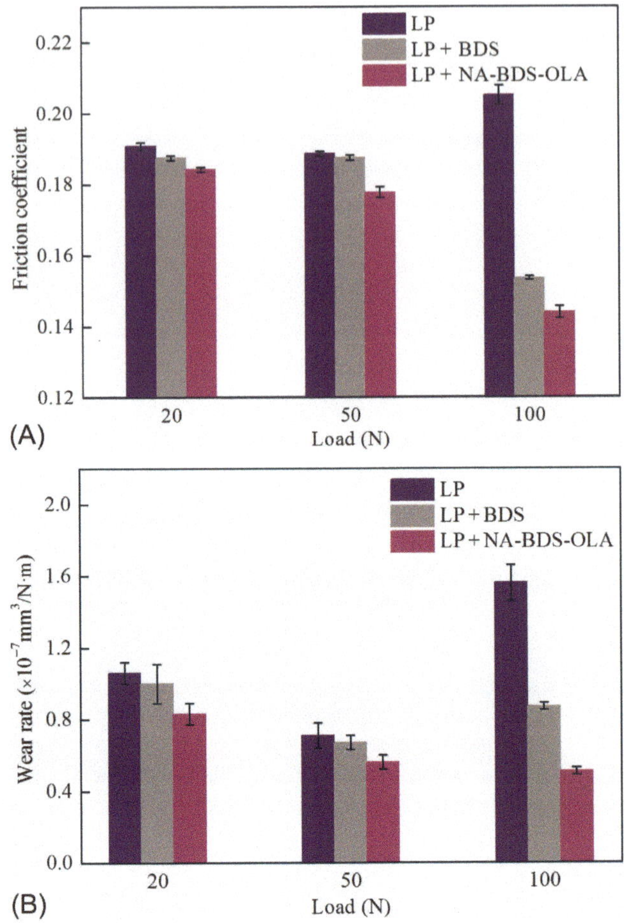

Fig. 6.19 Evolutions of the (A) average friction coefficient and (B) wear rates for lubrication using BDS and NA-BDS-OLA in LP at different loads (soot content = 0.3 wt%, speed = 50 mm/s, duration = 30 min).

Interestingly, at a low load (20 N, 50 N), the lubricating property of LP was slightly improved with the addition of BDS and NA-BDS-OLA. However, at a high load (100 N), the tribological properties of LP were significantly improved by adding BDS and NA-BDS-OLA. These results imply that an LP lubricant film forms at a low load and is easily damaged at a high load [9], possibly owing to the increased friction and wear of the LP under a high load. BDS and NA-BDS-OLA can play a positive role in protecting the friction interfaces [57]; hence, they have favorable antifriction and antiwear effects, improving the LP tribological properties under a high load.

Fig. 6.20 shows FESEM images, optical micrographs, and the corresponding worn surface profiles of wear scars on upper balls lubricated using LP, LP + BDS, and LP + NA-BDS-OLA. Compared with the FESEM images in Fig. 6.20A–C, the worn surfaces lubricated with LP exhibited many clear spalling pits. Numerous serious furrows and small pits were observed on the worn surface in the case of LP + BDS. For LP + NA-BDS-OLA, the worn surface exhibited no spalling pits and shallower furrows. The

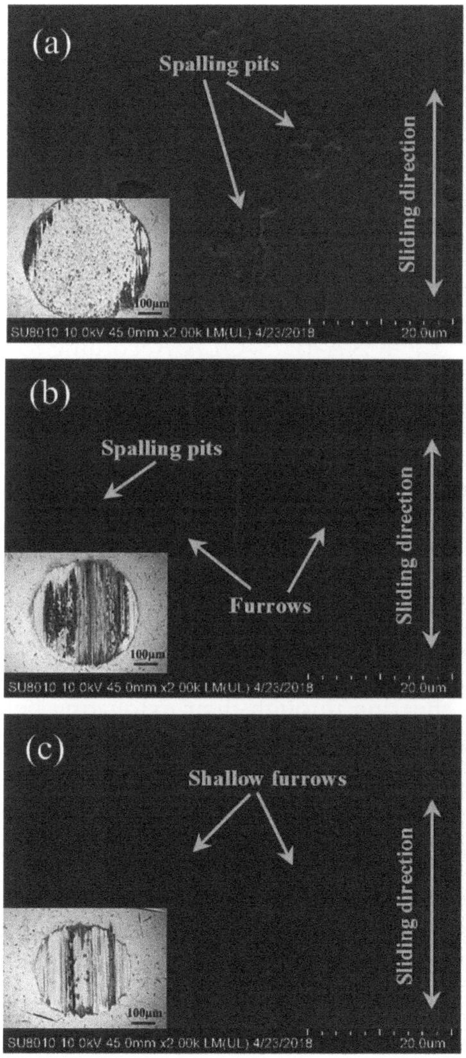

Fig. 4.20 See figure legend on next page.

(Continued)

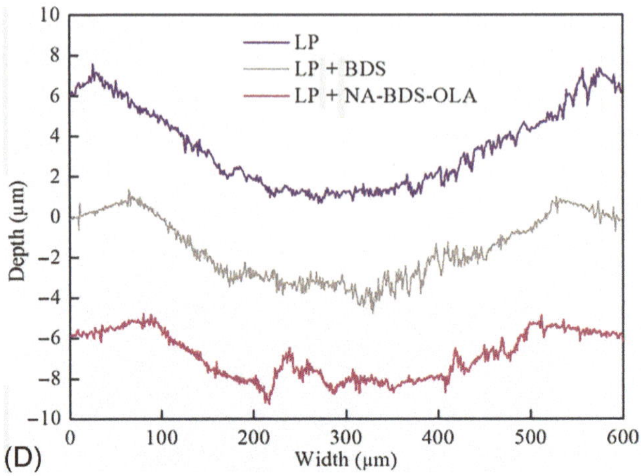

(D)

Fig. 6.20 FESEM images and optical micrographs of wear scars on upper balls lubricated with (A) LP, (B) LP + BDS, and (C) LP + NA-BDS-OLA. (D) Corresponding worn surface profiles (soot content = 0.3 wt%, load = 100 N, speed = 50 mm/s, duration = 30 min).

surface roughness *Ra* of the worn surface decreased in the following order: LP > LP + BDS > LP + NA-BDS-OLA. According to the optical micrographs in Fig. 6.20A–C and the corresponding worn surface profiles of the wear scars in Fig. 6.20D, the width and depth of the wear scars decreased in the following order: LP > LP + BDS > LP + NA-BDS-OLA. This indicates that the main type of wear in the case of lubrication with LP was fatigue wear [58], whereas for LP + BDS, the fatigue wear was reduced. However, abrasive wear was observed on the worn surface. Because BDS can fill the friction pairs and act as a ball bearing, the aggregation of BDS can cause abrasive wear [59, 60]. Apart from the spacing and ball-bearing effect, the carbon onion-like nanoparticles (NA-BDS-OLA) possibly alleviated the friction shearing through their exfoliating behavior [25], which is in accordance with the results shown in Fig. 6.19.

6.2.6 Lubrication mechanisms

Raman spectroscopy can be employed to obtain detailed information regarding the level of order in the structure by investigating the D and G peaks. The correlation between the structural changes of soot and Raman spectroscopic parameters has been discussed in the literature [27, 61]. Therefore, the order of soot before and after the friction test must be considered in Raman spectroscopic analyses. To investigate the differences in the friction reduction and wear resistance between BDS and NA-BDS, and between BDS and NA-BDS-OLA, we performed a Raman spectroscopic analysis

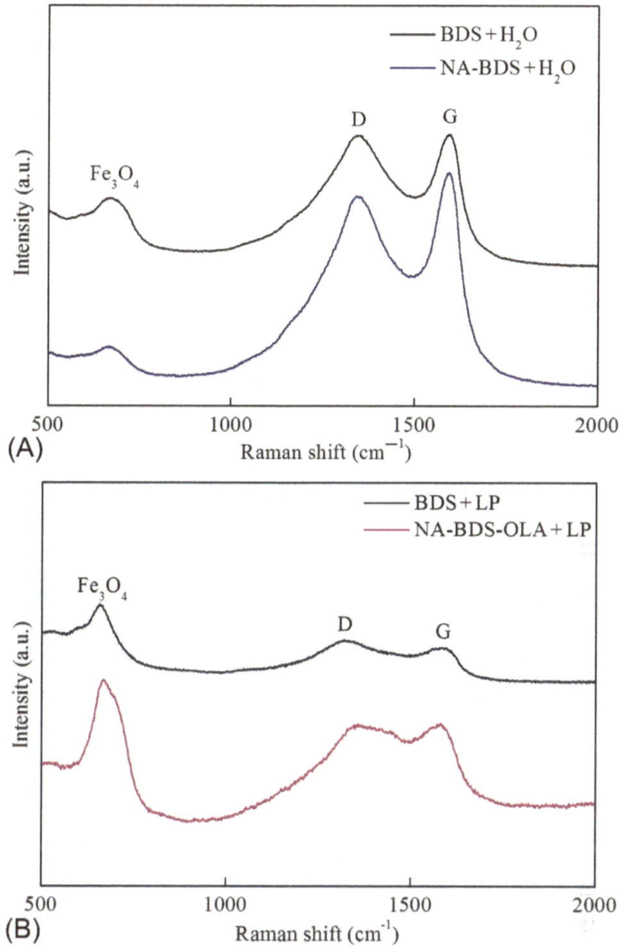

Fig. 6.21 Raman spectra of the worn surfaces of upper balls lubricated using (A) BDS + H_2O and NA-BDS + H_2O, and (B) BDS + LP and NA-BDS-OLA + LP.

of the wear surfaces. The Raman spectra of the worn surfaces of the upper balls are shown in Fig. 6.21.

The characteristic spectra of the D and G peaks were observed for the worn surfaces, indicating that the carbon nanoparticles deposited on the sliding surfaces formed a protective film. According to the Raman results in Figs. 6.10 and 6.21, compared with the I_D/I_G ratio for BDS (2.956), that of the worn surface lubricated using BDS in H_2O (2.986) was not significantly increased, whereas that of the worn surface lubricated using NA-BDS in H_2O (2.898) was significantly higher than that for NA-BDS (2.546).

The peak at 660/cm is attributed to Fe_3O_4, which formed a protective tribofilm through tribochemical reactions [62, 63]. Moreover, the I_D/I_G ratio for BDS (2.956) was similar to that of the worn surface lubricated using BDS in LP (2.927). However, the I_D/I_G ratio for the worn surface lubricated using NA-BDS-OLA in LP was 2.700, which was higher than that for NA-BDS-OLA (2.559). An Fe_3O_4 (660/cm) tribofilm was created during the sliding [64]. These results imply that the Fe_3O_4 formed is capable of acting as a lubricious oxide. In contrast to BDS, NA-BDS and NA-BDS-OLA were prone to being exfoliated, generating isolated graphitic layers under a high load, which reduced the friction and wear [25].

According to the foregoing analysis, BDS can provide spacing and act as a ball bearing between the rubbed surfaces [65–67]. However, NA-BDS and NA-BDS-OLA are structurally similar to onion-like carbon and have significantly higher degrees of graphitization order than BDS. The lubrication mechanisms of NA-BDS as a lubricant additive to H_2O and NA-BDS-OLA as a lubricant additive to LP can be summarized as follows. First, both types of soot can penetrate into the friction pairs and be deposited on the sliding surface, thereby playing a positive role in reducing the direct contact between the microconvex bodies during the friction experiment [59]. Second, both types of soot can act as a ball bearing between the friction pairs [10, 21]. Finally, under a high load, the graphitic layers of both types of soot are prone to being exfoliated and adsorbed onto the friction interface, resulting in easy sliding and a good lubrication effect between the surfaces [46, 57].

6.2.7 Conclusions

NA-BDS treated with nitric acid and NA-BDS-OLA modified with OLA were used as lubricant additives to H_2O and LP, respectively. Raman spectroscopy and FETEM indicated that both types of soot (NA-BDS and NA-BDS-OLA) exhibited a better order than BDS, having onion-like structures with a size of 35–40 nm. FTIR and particle-size distribution analysis revealed that NA-BDS and NA-BDS-OLA introduced strong hydrophilic and lipophilic groups, respectively, which reduced the average particle size.

To evaluate the tribological performance of the carbon onion-like nanoparticles derived from BDS, the tribological properties of NA-BDS in H_2O and NA-BDS-OLA in LP were investigated. We found that the carbon onion-like structure of the two types of soot reduced the friction coefficients and wear rates. The tribological mechanisms can be divided into three types: (1) a spacing effect, in which both types of soot act as spacers between the contact surfaces; (2) a ball-bearing effect, in which both types of soot create a rolling between the friction pairs; and (3) an exfoliation effect, in which a high load contributes to the formation of exfoliated graphitic layers from the two types of soot during sliding. These graphitic layers can be adsorbed onto the friction surfaces, resulting in easy sliding.

6.3 Lubricity of modified biodiesel soot in water

6.3.1 Introduction

With increasing energy supply shortages and the continuous improvements in environmental protection emission standards, research and development of green and renewable alternative fuels for automobile engines have received much attention. As discussed earlier, biodiesel is one of the current research hotspots in this field [68–70]. Biodiesel refers to a liquid fuel produced by the conversion of animal and vegetable oil by transesterification. Compared with fossil fuels, biodiesel has many advantages, including wide-ranging sources of raw materials, good biodegradability, excellent combustion performance, and fewer emissions of harmful substances; thus biodiesel is undoubtedly a new fuel with excellent development and application potential [71–74].

In the process of engine operation, under the condition of insufficient mixture of fuel and air, the generation of soot cannot be avoided [75]. Part of the soot is discharged to the atmosphere through the exhaust gas, causing serious harm to human health and the atmospheric environment [76]; another part of the soot enters the engine crankcase and can be found in the engine lubricating oil in the form of solid pollutants [77]. Salehi [78] and Kontau et al. [79] found that soot can wipe off the friction reaction film formed in the friction process of engine lubricating oil, and the formation rate of the friction reaction film is an important factor affecting the "corrosion wear" rate. Xiong et al. [80] found that soot adsorbs organomolybdenum additives in engine lubricating oil, which decreases the friction reduction performance of organic molybdenum. This shows that soot has a negative effect on the lubricating performance of engine lubricating oil. However, Hu et al. [81] found that in the presence of dispersant, a small amount of soot can improve the antifriction performance of base oil. Guo et al. [82] found that adding 0.01% diesel soot to PAO4 base oil can significantly improve the antiwear and antifriction performance of the PAO4 base oil. The results show that the proper amount of soot can play a role in lubrication, so soot has the potential to be a solid lubricant additive.

Although the oxygen content of biodiesel is higher than that of petrochemical diesel, which can reduce the generation of soot, in actual use, biodiesel will also burn incompletely to produce soot (BDS) [83]. With the rapid growth of the global biodiesel industry and the continuous increase of biodiesel production [80, 83], the effective utilization of biodiesel soot has become an urgent problem. As is commonly known, compared with oil-based lubricants, water-based lubricants have many advantages, including low cost, environmental friendliness, and wide-ranging sources, and they have broad application prospects in the field of industrial lubrication [83]. In order to meet the requirements of actual use conditions and environmental protection, it is necessary to use lubricating additives with excellent performance and green environmental protection.

Nanomaterials have many unique structural characteristics, which endow them with special physical and chemical properties that differ from traditional materials. Among these, the electrical, magnetic, optical, and mechanical properties are the most remarkable, and they have very important applications and development value. Nanomaterials have the characteristics of large specific surface areas, high diffusivities, easy sintering ability, and low melting points. When they are used as additives in the lubrication field, they can significantly improve the antiwear and extreme pressure properties of lubricants. Therefore, the research on nanomaterials as lubricating additives has become a focus worldwide.

When nanomaterials are used as lubrication additives, their action modes are different from those of traditional additives. Traditional lubricant additives are unable to repair a worn surface, which greatly limits the service life of the lubricated components. Due to the unique small-size effect and surface-interface effect of nanomaterials, compared with bulk materials, nanomaterials have lower melting points, along with higher adsorption capacities and reaction activities. Therefore, nanomaterials can be used to easily form a layer of low-shear film on the surface of a friction pair, so as to prevent direct contact between the pair, which can reduce the friction coefficient, reduce the wear volume, and repair the surface, to a certain extent. At present, carbon-based solid lubricating materials are a hot topic in the field of nanomaterials.

Carbon-based solid lubricating materials are very suitably used as water-based lubricating additives due to their advantages of nontoxicity and chemical stability [84]. Among them, onion-like carbon is a keen research focus in the field of carbon-based solid lubricant additives. Carbon onions show excellent lubrication performance, owing to their spherical structure and easy-to-peel property due to shear forces that manifest during the friction process [85]. However, current methods of preparing onion-like carbon (arc method, vapor deposition method, etc.) have the problems of low efficiency and high cost [86, 87]. Therefore, using a simple, efficient, and low-cost method to prepare carbon onions is a prerequisite for their wide application as a lubricating material.

Onion-like carbon nanoparticles are spherical in structure and roll between friction interfaces, thus playing the role of a "microbearing"; at the same time, due to the weak van der Waals force between the layers, the outer layers of the particles are easily peeled off to produce graphite sheets during the friction process, which is conducive to further improving lubrication performance [88–90]. Tevet et al. [91] used an AFM (atomic force microscopy) probe to apply pressure and a shear force to single, onion-like nanoparticles placed on a silicon wafer, and their test results verified the lubrication mechanism of rolling under low loads and peeling off under high loads.

At present, as a typical by-product of combustion, the soot produced by candle combustion has aroused research interest [92–95]. For example, Singh et al. [96] studied the separation and purification process of candle soot and the tribological properties of purified candle soot in engine lubricating oil. Their results showed that purified candle soot can effectively improve the antiwear performance of engine lubricating oil as an additive. Wei et al. [9, 10] found that, after high-temperature treatment or acidification, the degree of graphitization of candle soot increased significantly, and the treated candle soot showed an onion-like structure, which had obvious antifriction and antiwear effects in water. After further modifying the candle soot with cetyltrimethylammonium bromide, it was found that the modified candle soot could significantly improve the lubricating performance when used as an additive to liquid paraffin. Huang et al. [85] used heat treatment and chemical modification to obtain fluorinated candle soot and investigated its tribological properties as a lubricating additive to perfluoropolyether. Their results showed that the fluorinated candle soot had good antifriction and antiwear effects in the perfluoropolyether, which were attributed to the onion-like structure of the fluorinated candle soot, which played the role of rolling bearings during the friction process. The soot also filled in corrosion pits, which avoided further corrosion of the lubricated parts. In addition, the peeled graphite sheets adsorbed onto the friction surface to form a carbon lubricating film.

Hirata et al. [12] used diamond clusters and diamond particles as carbon sources, which were heated to 1730°C in an infrared radiation furnace under the protection of atmospheric-pressure argon to obtain carbon nano-onions. High-resolution transmission electron microscopy (HRTEM) showed that the carbon nano-onions obtained from the diamond clusters had a spherical multilayer structure, with a particle size of about 5–10 nm. However, carbon nano-onions with larger particle sizes were obtained from the diamond nanoparticles. The tribological properties of the carbon nano-onions were verified by a ball disk tester. The results showed that the friction coefficient of carbon nano-onions located between a silicon wafer and steel balls was stable, and lower than 0.1. Moreover, the wear rate of steel balls in the presence of carbon nano-onions was much lower than that in the presence of graphite powder. In general, carbon nano-onions with larger particle sizes have a better antifriction effect. The lubrication mechanism of carbon nano-onions is related to the surface roughness of the considered friction pair. When the roughness of the friction pair is large, the carbon nano-onions easily roll along the surface of the friction pair because of their weak adsorption capacity, thus reducing friction and wear. When the roughness of the friction pair is small, the carbon nano-onions form a lubricating layer, which aids the effect of lubrication and antiwear. It was also shown, however, that outer layer defects located on the surface of the carbon nano-onions could bond with the surface of the friction pair, which led to an increase of wear. Therefore, the higher the crystallinity of the carbon nano-onions, the better the lubrication and friction [12]. Next, Bucholz et al. [97] used dynamic simulations to confirm that carbon nano-onions undergo a process of transition from rolling to sliding under the action of increasing pressure and shear force. They showed that the carbon nano-onions were able to maintain their structure, and they played the role of a rolling bearing under low loads; under high loads, the structure peeled off and adsorbed under the action of increased shear force, thus forming a lubricating film and playing the role of lubrication by sliding.

Previous studies have found that biodiesel soot (BDS) has a higher graphitization degree than diesel soot particles, and it shows a certain onion-like carbon structure [98]. Therefore, BDS has the unique advantage of producing onion-like carbon. However, the preparation and application of onion-like carbon using BDS as a raw material has not been reported. Therefore, in this work, BDS was prepared by the thermal oxidation method (TO-BDS), and its morphology, composition, and dispersion were characterized by a variety of analytical methods. The tribological properties of pure water (H_2O) containing TO-BDS was investigated using a ball disk reciprocating friction tester, and the results were compared with those obtained for just H_2O.

The results of this work are not only beneficial for broadening the range of raw materials that can be used to prepare onion-like carbon and reduce the preparation costs of manufacturing onion-like carbon, but also for "turning waste into treasure" by promoting the recycling of BDS.

6.3.2 Experiments

6.3.2.1 Materials and preparation

The materials used were as follows: BDS collected using a self-made soot-capturing device [98]; ultrapure water obtained using a UPT-11-10T water purification system (Sichuan Ulupure Super Pure Technology Co., Ltd.); ball disc friction pair, with upper ball (ϕ 6 mm, roughness R_{a1} = 0.02 μm) and lower disk (50 × 50 × 5 mm, roughness R_{a2} = 0.5 μm) made of GCr15 bearing steel with hardness of 62-65 HRC.

The tribo-tests were performed with a ball-on-disc reciprocating tribometer (CFT-I, Lanzhou Zhongke Kaihua Technology Development Co. Ltd.); JEM-2100F field emission transmission electron microscope (FETEM, Japan Electronics Co., Ltd.); ESCALAB250Xi X-ray photoelectron spectrometer (XPS, Thermo Fisher Scientific Co., Ltd., United States); Nano-ZS90 zeta potential meter (Malvern Co., Ltd., United Kingdom); VK-X100 3D laser scanning microscope (Keyence Co., Japan); SU8020 field emission scanning electron microscope (FESEM, Hitachi Company, Japan); SL 200KS optical contact angle/interface tension meter (Solon Information Technology Co., Ltd., Shanghai); Raman spectrometer (Horiba Jobin Yvon Co., Ltd., France).

The preparation of TO-BDS was carried out as follows. The BDS was placed in an OTF-1200X tubular furnace from Hefei Kejing Materials Technology Co., Ltd., with constant air volume of 40 mL/min and oxidation of 2 h at 500°C. After natural cooling, it was placed in a drying dish for standby. The H_2O + TO-BDS was prepared by adding 0.05%, 0.1%, 0.2% and 0.3% TO-BDS to H_2O, stirring with a glass rod for 20 min, and dispersing it by ultrasonics for 30 min.

6.3.2.2 Tribological tests and characterization

Before starting the friction test, H_2O or H_2O + TO-BDS drops were added to the ball disk contact area. The friction test conditions were sliding distance 5 mm, test time 30 min, sliding speed 50 mm/s, and load 20, 50, 100 N, respectively. The tests were repeated three times for each group of tests and the average value of the three times was taken. The friction coefficient was recorded in time by the testing machine software; the wear scar diameter (d) of the upper ball was measured by the VK-X100 3D laser scanning microscope; and the wear volume (V) was calculated by the following formula [99]:

$$V = \frac{\pi h}{6}\left(\frac{3d^2}{4} + h^2\right), \tag{6.13}$$

$$h = R - \sqrt{R^2 - \frac{d^2}{4}}, \tag{6.14}$$

where d is diameter of the wear scar of upper ball, μm; R is radius of upper ball, μm; h is wear height of upper ball, μm.

The size and structure of BDS and TO-BDS were characterized using FETEM; the elemental valence and content of BDS and TO-BDS were characterized by XPS; the zeta potential and particle size of BDS and TO-BDS in H_2O were characterized by a Nano-ZS90 zeta potential instrument; the three-dimensional image of the ball milling mark area was characterized by three-dimensional laser scanning microscopy; and the morphology of the upper ball worn surfaces was characterized by FESEM. The morphology of H_2O and H_2O + TO-BDS drops on the surface of the lower disks was photographed with an SL 200KS optical contact angle/interface tensiometer, and the contact angle of the samples on the surface of the lower disks was measured. The optical micrograph of the worn surfaces of the lower disks, the Raman spectra of BDS, TO-BDS and the worn surface of the lower disks was obtained using an HR evolution Raman spectrometer.

6.3.3 Experiments

6.3.3.1 Characterization and analysis of soot

Fig. 6.22 shows the FETEM of BDS and TO-BDS. It can be seen from Fig. 6.22A and B that the aggregates of BDS and TO-BDS are composed of many near-spherical primary particles with chain structure, and the average particle size of the two kinds of soot is about 35 nm. It can be seen from Fig. 6.22C and D that the primary particle outer layers of BDS and TO-BDS have a discontinuous graphite sheet structure, but compared with BDS, TO-BDS has more of the graphite sheet structure and features of high bending and concentric orientation. This shows that heat oxidation treatment can improve the graphitization degree of BDS. This is similar to the results of Wei et al. [10], who found that, compared with untreated candle ash, treated candle ash in air atmosphere and at 450°C had a higher graphitization degree.

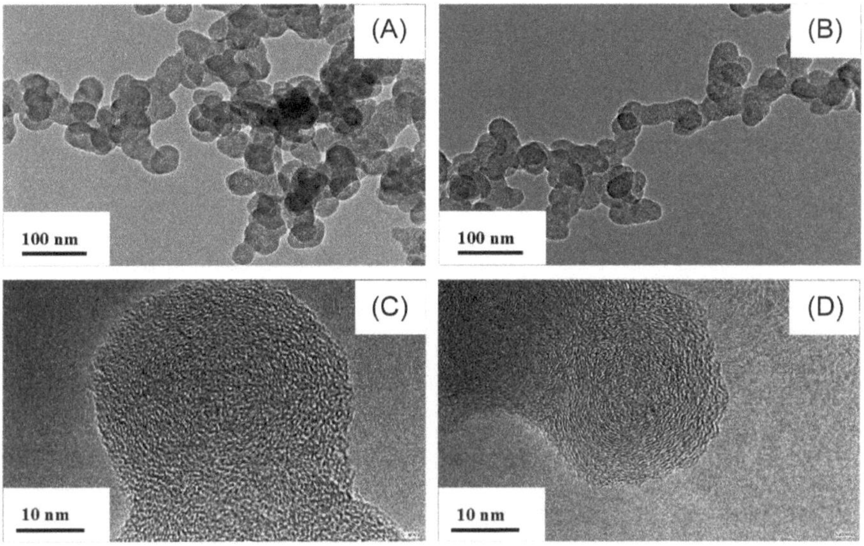

Fig. 6.22 FETEM images of BDS and TO-BDS: (A and C) FETEM image of BDS; (B and D) FETEM image of TO-BDS.

Fig. 6.23 shows the Raman spectra of BDS and TO-BDS. It can be seen from Fig. 6.23 that there are two characteristic peaks in the Raman spectra: the D peak (1337–1345/cm, a disordered peak) and the G peak (1585–1593/cm, a graphitization peak). The intensity ratio of the D peak to the G peak (I_D/I_G) can reflect the disorder degree of the soot, where a smaller I_D/I_G value indicates a lower degree of disorder and graphitization of the corresponding soot. The results show that I_D/I_G of TO-BDS (2.712) is lower than that of BDS (3.010). Therefore, the graphitization degree of TO-BDS is higher than that of BDS. This is consistent with the observed structures of BDS and TO-BDS in the FETEM images (Fig. 6.22).

According to the XPS test results, the surfaces of BDS and TO-BDS mainly contain C and O: the C and O content of the BDS surface is 92.21% and 7.79%, respectively, while the C and O content of the TO-BDS surface is 82.43% and 17.57%, respectively. This shows that the surface oxygen content of TO-BDS is higher than that of BDS. Furthermore, XPS spectra of BDS and TO-BDS were obtained via the XPS software, as shown in Fig. 6.24. It can be seen from Fig. 6.24A and C that the peaks at 284.8 eV can be attributed to the C–C bond, while the peaks at 286.5–286.9 eV and 288.8 eV belong to the C–O–C and C=O bonds, respectively. Compared with Fig. 6.24, it can be found that the C–C bond and C–O–C bond appear in the C_{1s} spectra of BDS and TO-BDS; in addition, the C=O

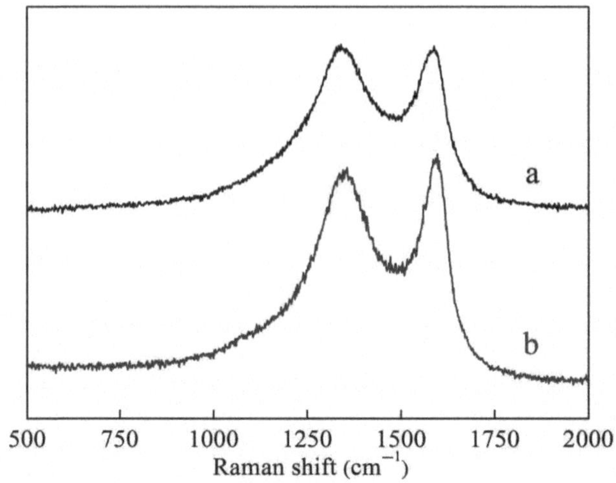

Fig. 6.23 Raman spectra of BDS (A) and TO-BDS (B).

bond appears in the C_{1s} spectrum of TO-BDS. These results indicate that the C=O bond has been introduced into TO-BDS [100] after the heat-oxidation treatment. It can also be seen from Fig. 6.24B and D that the peaks at 531.6–531.8 eV belong to either C–O–C or C=O bonds, while the peaks at 533.6–533.7 eV can be attributed to C–OH bonds. As shown in Table 6.2, C–O–C, C=O bonds and C–OH bonds can be seen in the O_{1s} spectra of both BDS and TO-BDS, and the content of oxygen-containing groups in TO-BDS is greater than that in BDS. This indicates that the TO-BDS surface contains more hydrophilic groups, suggesting that TO-BDS can more easily disperse in H_2O than BDS [101].

Table 6.3 shows the average particle sizes and zeta potentials of BDS and TO-BDS in H_2O. It can be seen from Table 6.3 that both the BDS and TO-BDS surfaces have negative charges, but the negative charges on the TO-BDS surface are greater than those on the BDS. An increased negative charge on a particle surface is conducive to improving its dispersion in H_2O; hence, the dispersion of TO-BDS in H_2O is better than that of BDS. This additional charge may be due to the presence of oxygen-containing groups on the surfaces of BDS and TO-BDS, with more being present on the surface of TO-BDS relative to BDS. In addition, the average particle size of the TO-BDS aggregate is smaller than that of the BDS aggregate. It can be seen that the average particle size of BDS in H_2O decreased due to the heat-oxidation treatment. This shows that thermal-oxidation treatment can improve the dispersion effect of BDS in H_2O.

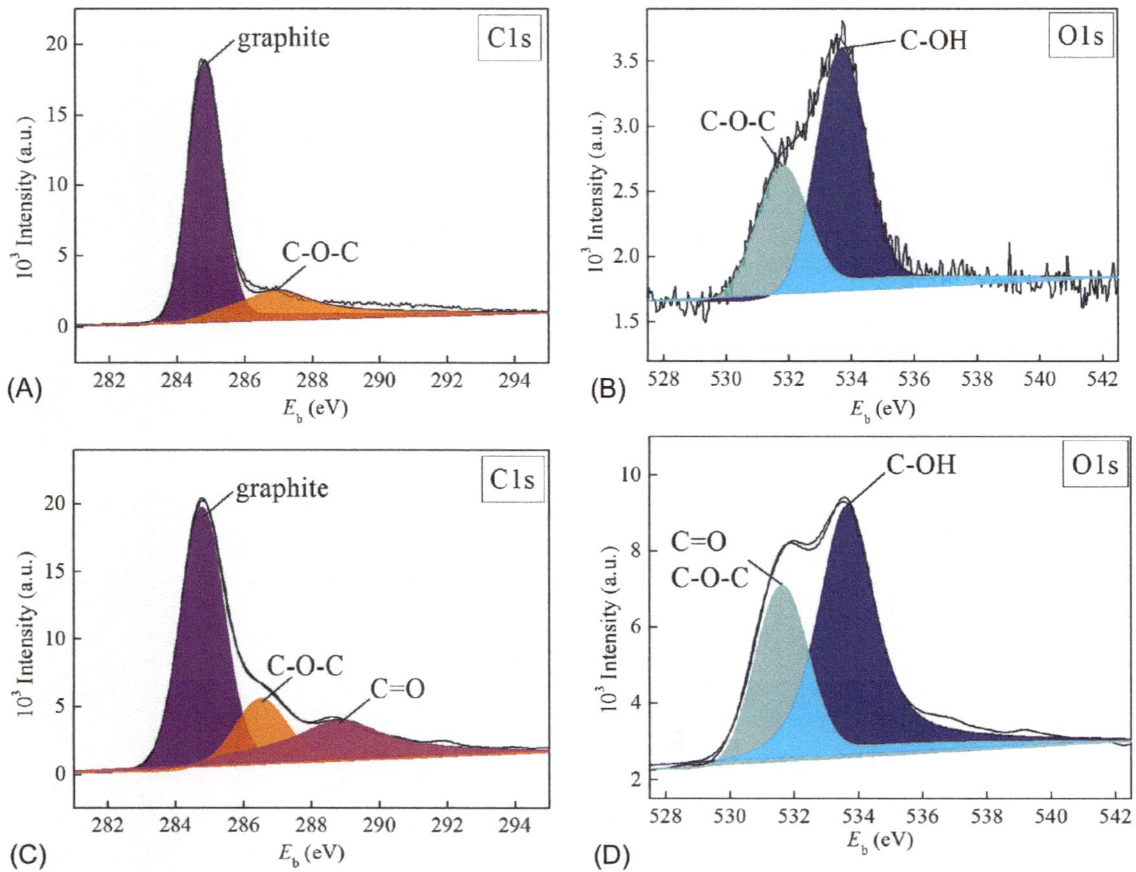

Fig. 6.24 XPS spectra of BDS and TO-BDS: (A) C_{1s} spectra of BDS; (B) O_{1s} spectra of BDS; (C) C_{1s} spectra of TO-BDS; (D) O_{1s} spectra of TO-BDS.

Table 6.2 Oxygen groups content on BDS and TO-BDS surface.

Soot sample	Oxygen groups atom content (%)	
	O (C–OH)	O (C–O–C, C=O)
BDS	4.90	2.89
TO-BDS	11.43	6.14

Table 6.3 Zeta potential and average particle size of BDS and TO-BDS in water.

Soot sample	Zeta potential (mV)	Average particle size (nm)
BDS	−5.07	213.2
TO-BDS	−17.6	191.6

In our experiments, we added BDS and TO-BDS into H_2O according to a dosage mass fraction of 0.2% for each. The solution was stirred for 20 min and dispersed by ultrasonic treatment for 30 min. Photos of the BDS and TO-BDS dispersed in water were then taken, as shown in Fig. 6.25. It can be seen from Fig. 6.25A that, after agitation and ultrasound treatment, part of the BDS dispersed in H_2O, but some of the BDS was still attached to the walls of the sample bottle; it was also found that BDS was attached to the walls at the bottom of the sample bottle through inverted observation. It can be seen from Fig. 6.25B that, after agitation and ultrasound treatment, the TO-BDS had completely dispersed in the H_2O and no TO-BDS had adhered to the bottle walls before or after inversion. Compared with BDS, TO-BDS is more easily dispersed in H_2O (the XPS, zeta potential in H_2O, and average particle size results all confirm this), which is conducive

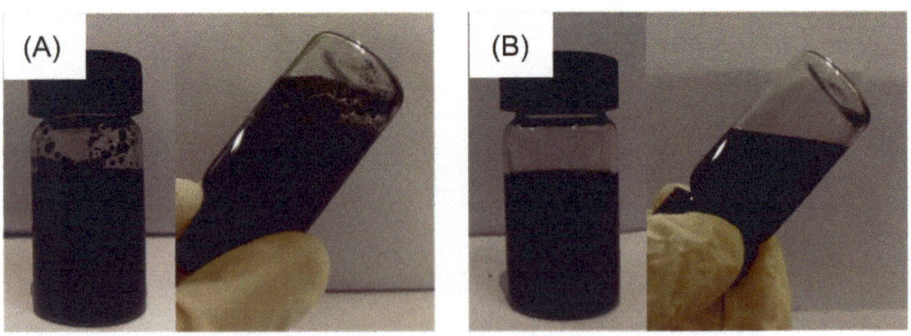

Fig. 6.25 Photographs of BDS and TO-BDS dispersed in water: (A) BDS + H_2O; (B) TO-BDS + H_2O.

to the TO-BDS behaving at a friction interface as lubricating particles [102]. Therefore, TO-BDS is more suitable as a lubricating additive to H_2O than BDS.

6.3.3.2 Friction and wear properties of H_2O containing TO-BDS

Fig. 6.26 shows the average friction coefficient and wear volume when TO-BDS is added to water under the conditions of a 50-N load, a sliding speed of 50 mm/s, and a test time of 30 min. It can be seen from Fig. 6.26A that, with an increase of the TO-BDS mass fraction (from 0%–0.2%), the average friction coefficient of the ball disk friction pair decreases significantly, which shows that adding TO-BDS can improve the friction-reduction characteristics of H_2O. When the TO-BDS mass fraction is 0.2%, the average friction coefficient decreases from 0.356 to 0.242, which is a decrease of 32.0%; further increasing the TO-BDS mass fraction does not change the average friction coefficient very much. It can be seen from Fig. 6.26B that when TO-BDS is added to H_2O, the wear volume of the upper ball significantly decreases, indicating that TO-BDS can effectively enhance the antiwear characteristics of H_2O. Overall, when the mass fraction of TO-BDS increases from 0% to 0.2%, the wear volume shows a significant decreasing trend. The wear volume (2.3×10^6 μm^3) of H_2O + 0.2%TO-BDS is 60.0% lower than that of just H_2O (9.2×10^5 μm^3), and with further increase in the TO-BDS mass fraction the wear volume did not change significantly. In conclusion, TO-BDS can significantly improve the antifriction and antiwear characteristics of H_2O. This is similar to the results of Chen et al. [103], who demonstrated that water-soluble carbon-based materials can provide good antifriction and antiwear properties in water.

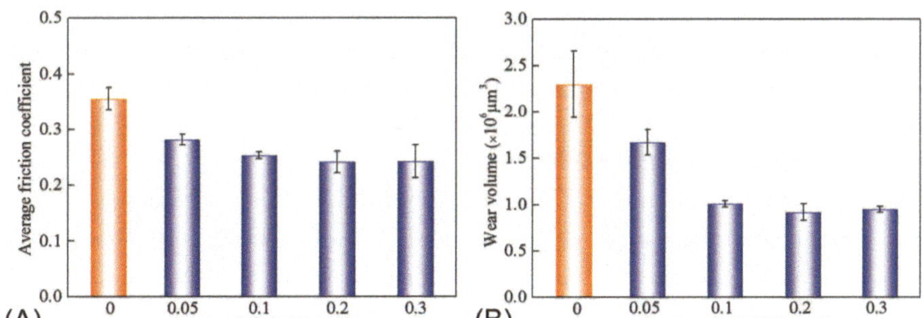

Fig. 6.26 Average friction coefficient and wear volumes of ball-on disk contact lubricated by water plus different content of TO-BDS: (A) Average friction coefficient; (B) Wear volume.

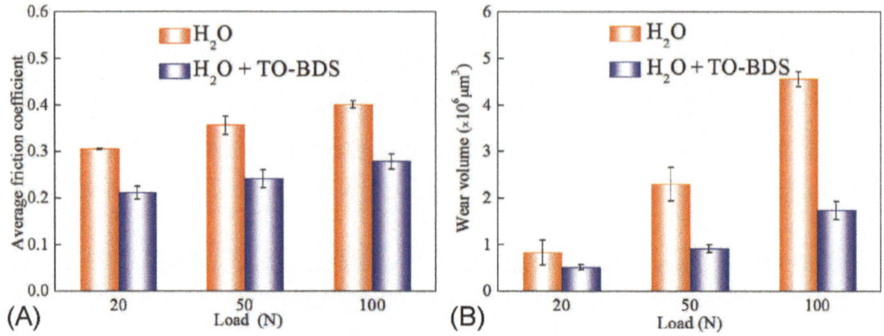

Fig. 6.27 Variations of average friction coefficient and wear volumes lubricated by water with or without TO-BDS at different loads: (A) Average friction coefficient; (B) Wear volume.

Fig. 6.27 shows the average friction coefficient and wear volume for different loads for H_2O and H_2O + TO-BDS under the conditions of a 0.2% TO-BDS mass fraction (latter case only), a 50-mm/s sliding speed, and a 30-min test time. According to Fig. 6.27A, the average friction coefficient under lubrication of H_2O + TO-BDS is lower than that under lubrication of H_2O only in the load range (20–100 N). When the load is 100 N, the average friction coefficient of H_2O + TO-BDS (0.278) is 30.8% lower than that of just H_2O (0.402), which again proves that the friction-reduction performance of H_2O+TO-BDS is better than that of just H_2O. It can be seen from Fig. 6.27B that the changing trend of the wear volume is similar to the changing trend seen for the average friction coefficient. When the load is 100 N, the wear volume of H_2O + TO-BDS (1.7×10^6 μm^3) is 63.0% lower than that without TO-BDS (4.6×10^6 μm^3). The results also show that H_2O + TO-BDS has better antiwear properties compared with just H_2O. In conclusion, TO-BDS can improve the tribological properties of H_2O across a wide range of loads. Even under a high load, TO-BDS can still enter the friction interface to play a lubricating role.

6.3.3.3 Analysis of wear surfaces

Fig. 6.28 shows 3D and FESEM images of the worn surfaces of upper balls under the lubrication of H_2O and H_2O+TO-BDS (test conditions: a TO-BDS mass fraction of 0.2%, a sliding speed of 50 mm/s, a test time of 30 min, and a load of 100 N). In Fig. 6.28A and B, it can be seen from the 3D image of the worn surfaces of the upper balls that the widths of the worn scars under H_2O lubrication are larger than those seen under H_2O + TO-BDS lubrication, suggesting that the wear of the upper ball under the

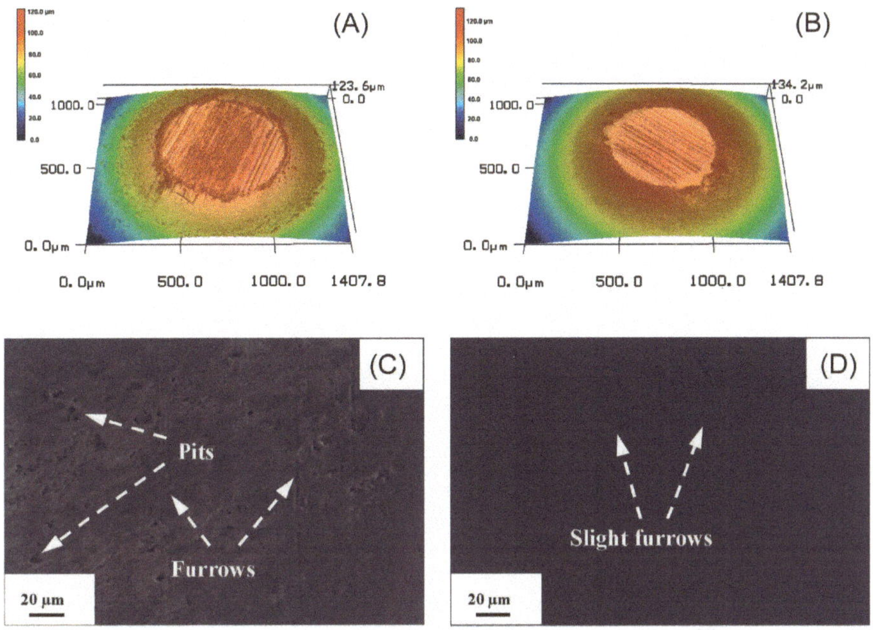

Fig. 6.28 3D topographies and FESEM images of upper ball worn surfaces lubricated by water with or without TO-BDS: (A) H_2O; (B) $H_2O + 0.2\%$TO-BDS.

former lubrication is more serious than that of the latter, which corresponds with the results shown in Fig. 6.27B. It can be seen from Fig. 6.28C and D that there is serious corrosion and wear in the abrasion-marked area of the H_2O friction test, with a large number of corrosion pits and furrows, while only slight furrows and no corrosion pits are found in the abrasion-marked area under the H_2O + TO-BDS lubrication. This is due to the fact that TO-BDS can enter into the friction-contact area and play the role of a rolling bearing at the friction interface. In addition, TO-BDS participates in the formation of a lubricating film, thus achieving the effect of preventing corrosion and wear [90, 104].

6.3.3.4 Analysis of the mechanism of friction reduction and antiwear of TO-BDS

The wettability of a liquid lubricant on a friction pair material is closely related to its corresponding tribological characteristics [104]. Therefore, the contact angle of H_2O and H_2O + TO-BDS on the lower disk surface was further investigated, as shown in Fig. 6.29. According to Fig. 6.29, the contact angles of the H_2O and H_2O + TO-BDS-coated lower disks are 98 and 77 degrees, respectively. The contact angle directly reflects the wettability of

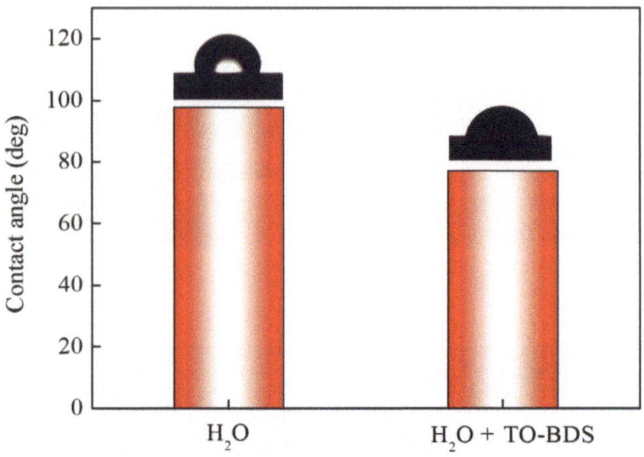

Fig. 6.29 The contact angle of H_2O and H_2O + 0.2%TO-BDS on lower disk surfaces.

the liquid lubricant on the friction pair material. The smaller the contact angle is, the easier the liquid lubricant will spread on the surface of the friction pair material [105]. Therefore, the wettability of H_2O + TO-BDS on the lower disk surface is better than that of just H_2O. This shows that H_2O + TO-BDS can more easily form a lubricating film on the surface of the friction pair than just H_2O, thus maintaining good antifriction and antiwear properties [106]. This conclusion is consistent with the results of the tribological test.

Fig. 6.30 shows an optical micrograph of the wear-mark area of the lower disks under lubrication of H_2O and H_2O + TO-BDS, and Raman spectrum analysis of the wear-mark area of the lower disk lubricated by H_2O and H_2O + TO-BDS (the test conditions are the same as in Fig. 6.28). In Fig. 6.30A and C, obvious corrosion traces can be seen in the wear-mark area of the lower disk coated with H_2O. A further Raman spectrum analysis was carried out on the wear-mark surface. According to the different shades in the optical micrograph of the wear-mark area of the lower disks, the optical micrograph can be divided into a black area and a white area. It can be seen that the black area (point 1) and white area (point 2) of the wear-mark surface show characteristic peaks attributed to Fe_3O_4 at 669/cm, which are the result of the oxide produced by the tribochemical reaction during the friction process under H_2O lubrication [107]. It can be seen from Fig. 6.30B and D that there is no obvious corrosion phenomenon in the wear-mark area of the lower disks lubricated with H_2O + TO-BDS. The black area (point 3) and white area (point 4) of the wear-mark surface

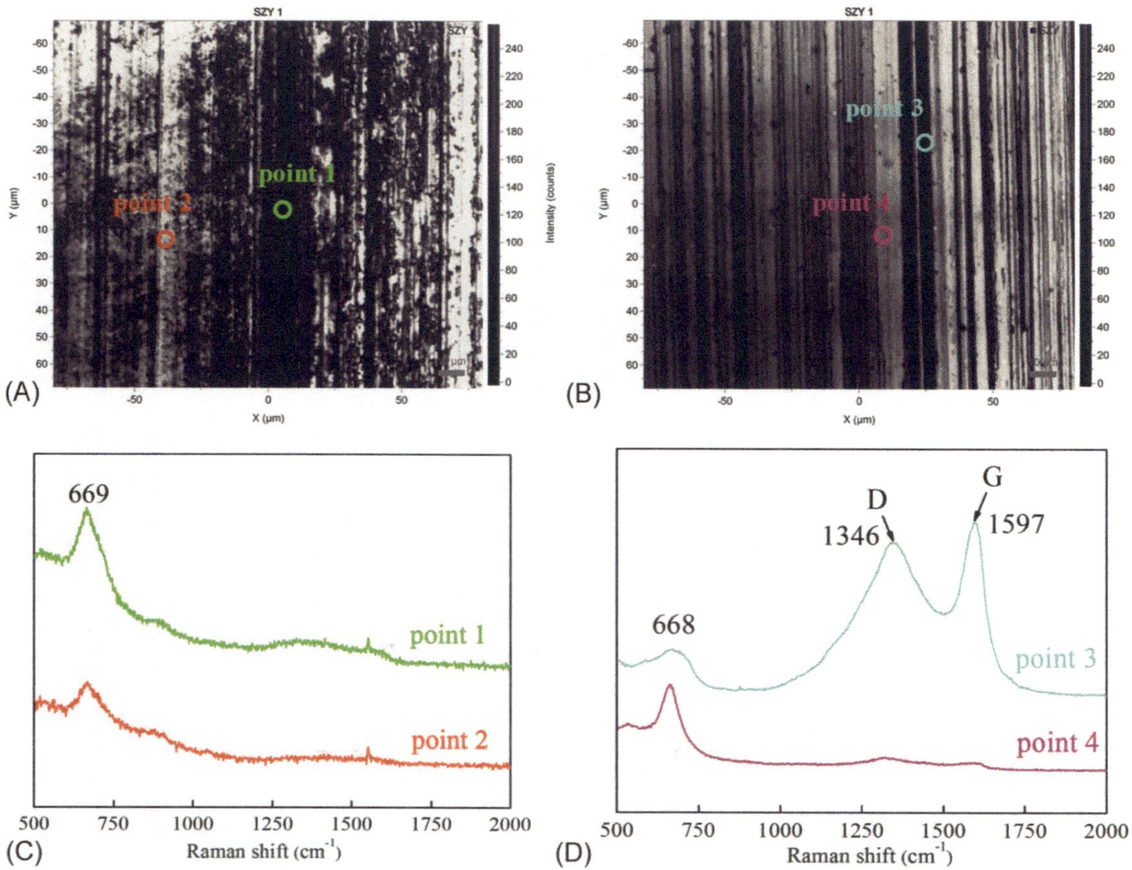

Fig. 6.30 Optical microscopy images of lower disks lubricated by water with or without TO-BDS; Raman spectra of TO-BDS and lower disk worn surfaces lubricated by H_2O and H_2O + TO-BDS: (A and C) H_2O; (B and D) H_2O + 0.2%TO-BDS.

have the characteristic peaks (668/cm) of Fe_3O_4. Meanwhile, the black area (point 3) has strong D and G peaks, while the white area (point 4) has weak D and G peaks. This shows that TO-BDS forms a lubricating film during the process of friction, and thus plays a protective role at the friction interface [108]. In addition, combined with the analysis shown in Fig. 6.23, it can be found that the I_D/I_G value of TO-BDS (2.712) is smaller than that of the black area (point 3, the I_D/I_G value of (2.928)), which means that the graphitization degree of TO-BDS changes before and after the friction test, where the friction induces the outer graphite sheet of TO-BDS to peel off, and the stripped graphite sheet slides at the friction interface, thus reducing friction and wear [27].

6.3.4 Conclusions

The conclusions of this section are summarized as follows:

(1) TO-BDS is a suitable water-based lubricant additive. Compared with BDS, TO-BDS has more oxygen-containing functional groups, more negative charges on its surface, smaller average particle size in H_2O, and a better dispersion effect.

(2) The average friction coefficient and wear volume of H_2O decreased by 30.8% and 63.0%, respectively, when TO-BDS was added at a 0.2% mass fraction under a 100-N load, indicating that TO-BDS can significantly improve the antifriction and antiwear performance of H_2O.

(3) During the friction process, TO-BDS plays the role of a rolling bearing in the friction-contact area. The wettability of H_2O+TO-BDS is better than that of just H_2O on the surface of a friction pair, where TO-BDS more easily forms a lubricating film due to the lubricating effect of the graphite sheet produced by TO-BDS peeling, which results in the effect of reducing friction and wear in H_2O.

References

[1] Alptekin E. Emission, injection and combustion characteristics of biodiesel and oxygenated fuel blends in a common rail diesel engine. Energy 2017;119:44–52.
[2] Lahane S, Subramanian KA. Effect of different percentages of biodiesel–diesel blends on injection, spray, combustion, performance, and emission characteristics of a diesel engine. Fuel 2015;139:537–45.
[3] Hu E, Hu X, Liu T, Fang L, Dearn KD, Xu H. The role of soot particles in the tribological behavior of engine lubricating oils. Wear 2013;304:152–61.
[4] Su P, Xiong Y, Liu X, Yang H. Morphology, structure and tribological behavior of diesel soot. Tribology 2017;37:83–9.
[5] Guo MF, Cai ZB, Zhang ZC, Zhu MH. Characterization and lubrication performance of diesel soot nanoparticles as oil lubricant additives. RSC Adv 2015;5:101965–74.

[6] Li C, Wei D, Zhuang Y, Song R, Hu X. Effect of biodiesel soot on tribological behavior of liquid paraffin. Chin Pet Process Pe Technol 2018;20:106–13.
[7] Merchan-Merchan W, Sanmiguel SG, McCollam S. Analysis of soot particles derived from biodiesels and diesel fuel air-flames. Fuel 2012;102:525–35.
[8] Lapuerta M, Oliva F, Agudelo JR, Boehman AL. Effect of fuel on the soot nanostructure and consequences on loading and regeneration of diesel particulate filters. Combust Flame 2012;159:844–53.
[9] Wei JX, Yao MH, Cai MR, Zhou F. Tribological properties of cetyltrimethyl ammonium bromide modified candle soot as effective lubricant additive in oil. Tribology 2014;34:428–36.
[10] Wei JX, Cai MR, Zhou F, Liu WM. Candle soot as particular lubricant additives. Tribol Lett 2014;53:521–31.
[11] Antusch S, Dienwiebel M, Nold E, Albers P, Spicher U, Scherge M. On the tribochemical action of engine soot. Wear 2010;269:1–12.
[12] Hirata A, Igarashi M, Kaito T. Study on solid lubricant properties of carbon onions produced by heat treatment of diamond clusters or particles. Tribol Int 2004;37:899–905.
[13] Li C, Hu XG. Green biomass ash catalyst for biodiesel synthesis. Acta Petrol Sin (Pet Process Sect) 2018;34:78–85.
[14] Mohanty B, Verma AK, Claesson P, Bohidar HB. Physical and anti-microbial characteristics of carbon nanoparticles prepared from lamp soot. Nanotechnology 2007;18:445102.
[15] Salehi FM, Khaemba DN, Morina A, Neville A. Corrosive–abrasive wear induced by soot in boundary lubrication regime. Tribol Lett 2016;63:19.
[16] Osswald S, Yushin G, Mochalin V, Kucheyev SO, Gogotsi Y. Control of sp^2/sp^3 carbon ratio and surface chemistry of nanodiamond powders by selective oxidation in air. J Am Chem Soc 2006;128:11635–42.
[17] Vander Wal RL, Yezerets A, Currier NW, Kim DH, Wang CM. HRTEM study of diesel soot collected from diesel particulate filters. Carbon 2007;45:70–7.
[18] Hu E, Hu X, Liu T, Liu Y, Song R, Chen Y. Investigation of morphology, structure and composition of biomass-oil soot particles. Appl Surf Sci 2013;270:596–603.
[19] Cain JP, Gassman PL, Wang H, Laskin A. Micro-FTIR study of soot chemical composition—evidence of aliphatic hydrocarbons on nascent soot surfaces. Phys Chem Chem Phys 2010;12:5206–18.
[20] Li CC, Huang CL. Preparation of clear colloidal solutions of detonation nanodiamond in organic solvents. Colloids Surf A 2010;353:52–6.
[21] Zhang ZC, Cai ZB, Peng JF, Zhu MH. Comparison of the tribology performance of nano-diesel soot and graphite particles as lubricant additives. J Phys D Appl Phys 2015;49: 045304.
[22] Lv J, Bai M, Cui W, Li X. The molecular dynamic simulation on impact and friction characters of nanofluids with many nanoparticles system. Nanoscale Res Lett 2011;6:200.
[23] Green DA, Lewis R. Effect of soot on oil properties and wear of engine components. J Phys D Appl Phys 2007;40:5488–501.
[24] Gong Z, Shi J, Zhang B, Zhang J. Graphene nano scrolls responding to superlow friction of amorphous carbon. Carbon 2017;116:310–7.
[25] Huang G, Yu Q, Ma Z, Cai M, Zhou F, Liu W. Fluorinated candle soot as the lubricant additive of perfluoropolyether. Tribol Lett 2017;65:28.
[26] Joly-Pottuz L, Matsumoto N, Kinoshita H, Vacher B, Belin M, Montagnac G, et al. Diamond-derived carbon onions as lubricant additives. Tribol Int 2008;41:69–78.
[27] Cao ZF, Xia YQ. Study on the preparation and tribological properties of fly ash as lubricant additive for steel/steel pair. Tribol Lett 2017;65:104.

[28] Su P, Xiong Y, Liu X, Yang H, Fan L, Liu P. The Role of diesel soot in the tribological behavior of 150SN base oil. Chin Pet Process Pe Technol 2017;19:89–95.
[29] Tevet O, Von-Huth P, Popovitz-Biro R, Rosentsveig R, Wagner HD, Tenne R. Friction mechanism of individual multilayered nanoparticles. Proc Natl Acad Sci 2011;108:19901–6.
[30] Li YR, Zhao J, Tang C, He Y, Wang Y, Chen J. Highly exfoliated reduced graphite oxide powders as efficient lubricant oil additives. Adv Mater Interfaces 2016;3:1600700.
[31] Singh V, Bux F, Sharma YC. A low cost one pot synthesis of biodiesel from waste frying oil (WFO) using a novel material, β-potassium dizirconate (β-$K_2Zr_2O_5$). Appl Energy 2016;172:23–33.
[32] Kohse-Höinghaus K, Oßwald P, Cool TA, Kasper T, Hansen N, Qi F, et al. Biofuel combustion chemistry: from ethanol to biodiesel. Angew Chem Int Ed 2010;49:3572–97.
[33] Lin BF, Huang JH, Huang DY. Experimental study of the effects of vegetable oil methyl ester on DI diesel engine performance characteristics and pollutant emissions. Fuel 2009;88:1779–85.
[34] Hu E, Hu X, Liu T, Song R, Dearn KD, Xu HM. Role of TiF_3 catalyst in the tribological properties of biofuel soot-contaminated liquid paraffin. Tribol Int 2014;77:122–31.
[35] Green DA, Lewis R, Dwyer-Joyce RS. Wear effects and mechanisms of soot-contaminated automotive lubricants. Proc Inst Mech Eng Part J J Eng Tribol 2006;220:159–69.
[36] Joly-Pottuz L, Vacher B, Ohmae N, Martin JM, Epicier T. Anti-wear and friction reducing mechanisms of carbon nano-onions as lubricant additives. Tribol Lett 2008;30:69–80.
[37] Jung Y, Hwang J, Bae C. Assessment of particulate matter in exhaust gas for biodiesel and diesel under conventional and low temperature combustion in a compression ignition engine. Fuel 2016;165:413–24.
[38] Savic N, Rahman MM, Miljevic B, Saathoff H, Naumann KH, Leisner T, et al. Influence of biodiesel fuel composition on the morphology and microstructure of particles emitted from diesel engines. Carbon 2016;104:179–89.
[39] Growney DJ, Mykhaylyk OO, Middlemiss L, Fielding LA, Derry MJ, Aragrag N, et al. Is carbon black a suitable model colloidal substrate for diesel soot. Langmuir 2015;31:10358–69.
[40] Kontou A, Southby M, Spikes HA. Effect of steel hardness on soot wear. Wear 2017;390:236–45.
[41] Hu E, Dearn KD, Yang BX, Song RH, Xu YF, Hu XG. Tribofilm formation and characterization of lubricating oils with biofuel soot and inorganic fluorides. Tribol Int 2017;107:163–72.
[42] Clague ADH, Donnet JB, Wang TK, Peng JCM. A comparison of diesel engine soot with carbon black. Carbon 1999;37:1553–65.
[43] Ferraro G, Fratini E, Rausa R, Fiaschi P, Baglioni P. Multiscale characterization of some commercial carbon blacks and diesel engine soot. Energy Fuel 2016;30:9859–66.
[44] Boehman AL, Song J, Alam M. Impact of biodiesel blending on diesel soot and the regeneration of particulate filters. Energy Fuel 2005;19:1857–64.
[45] Zhang Y, Zhang R, Rao L, Kim D, Kook S. The influence of a large methyl ester on in-flame soot particle structures in a small-bore diesel engine. Fuel 2017;194:423–35.
[46] Deng X, Mammen L, Butt HJ, Vollmer D. Candle soot as a template for a transparent robust superamphiphobic coating. Science 2012;335:67–70.
[47] Thomson M, Mitra T. A radical approach to soot formation. Science 2018;361:978–9.
[48] Mutyala KC, Wu YA, Erdemir A, Sumant AV. Graphene-MoS_2 ensembles to reduce friction and wear in DLC-steel contacts. Carbon 2019;146:524–7.

[49] Wang WZ, Huang P. Study on the lubrication state of frictional pairs with different surface roughness based on Stribeck curves. Tribology 2004;24:254–7.
[50] Xu YF, Liu ZC, Dearn KD, Dong YH, You T, Hu XG. Thermo-tribological behaviour of microgels for improved aqueous lubrication for steel/UHMWPE contact. Tribol Int 2019;130:63–73.
[51] Xu YF, Yu JY, Dong YH, You T, Hu XG. Boundary lubricating properties of black phosphorus nanosheets in polyalphaolefin oil. J Tribol 2019;141: 072101.
[52] Rosenkranz A, Freeman L, Fleischmann S, Lasserre F, Fainman Y, Talke FE. Tip-enhanced Raman spectroscopy studies of nanodiamonds and carbon onions. Carbon 2018;132:495–502.
[53] Du SN, Sun JL, Wu P. Preparation, characterization and lubrication performances of graphene oxide-TiO_2 nanofluid in rolling strips. Carbon 2018;140:338–51.
[54] Klokkenburg M, Hilhorst J, Erne BH. Surface analysis of magnetite nanoparticles in cyclohexane solutions of oleic acid and oleylamine. Vib Spectro 2007;43:243–8.
[55] Zhang GQ, Xu Y, Xiang XZ, Zheng GL, Zeng XQ, Li ZP, et al. Tribological performances of highly dispersed graphene oxide derivatives in vegetable oil. Tribol Int 2018;126:39–48.
[56] Xiao HP, Liu SH, Wang DG, Chen Y. Abrasion–corrosion behaviors of steel–steel contact in seawater containing abrasive particles. Tribol Trans 2018;61:12–8.
[57] Fan XQ, Li W, Fu HM, Zhu MH, Wang LP, Cai ZB, et al. Probing the function of solid nanoparticle structure under boundary lubrication. ACS Sustain Chem Eng 2017;5:4223–33.
[58] Liu YR, Hu KH, Hu EZ, Guo JH, Han CL, Hu XG. Double hollow MoS_2 nano-spheres: synthesis, tribological properties, and functional conversion from lubrication to photocatalysis. Appl Surf Sci 2017;392:1144–52.
[59] Alazemi AA, Etacheri V, Dysart AD, Stacke LE, Pol VG, Sadeghi F. Ultrasmooth submicrometer carbon spheres as lubricant additives for friction and wear reduction. ACS Sustain Chem Eng 2015;7:5514–21.
[60] Penchaliah R, Harvey TJ, Wood RJK, Nelson K, Powrie HEG. The effects of diesel contaminants on tribological performance on sliding steel on steel contacts. Proc Inst Mech Eng Part J J Eng Tribol 2011;225:779–97.
[61] Knauer M, Carrara M, Rothe D, Niessner R, Ivleva NP. Changes in structure and reactivity of soot during oxidation and gasification by oxygen, studied by micro-Raman spectroscopy and temperature programmed oxidation. Aerosol Sci Tech 2009;43:1–8.
[62] Zhang X, Niu YA, Meng XD, Li Y, Zhao JP. Structural evolution and characteristics of the phase transformations between α-Fe_2O_3, Fe_3O_4 and γ-Fe_2O_3 nanoparticles under reducing and oxidizing atmospheres. CrystEngComm 2013;15:8166–72.
[63] Zin V, Agresti F, Barison S, Litti L, Fedele L, Meneghetti M, et al. Effect of external magnetic field on tribological properties of goethite (a-FeOOH) based nanofluids. Tribol Int 2018;127:341–50.
[64] Shi B, Guo JH, Cao XA, Hu EZ, Hu KH. Effects of carbon soot from the combustion of diesel fuels on the tribological properties of lubricating oil and diesel fuels. Ind Lubr Tribol 2018;70:532–7.
[65] Zhang ZC, Cai ZB, Peng JF, Zhu MH. Comparison of the tribology performance of nano-diesel soot and graphite particles as lubricant additives. J Phys D Appl Phys 2015;49: 045304.
[66] Ma TB, Wang LF, Hu YZ, Li X, Wang H. A shear localization mechanism for lubricity of amorphous carbon materials. Sci Rep 2014;4:3662.
[67] Li C, Li ML, Wang XY, Feng WM, Zhang QQ, Wu B, Hu XG. Novel carbon nanoparticles derived from biodiesel soot as lubricant additives. Nanomater 2019;9:1115.

[68] Khandal SV, Banapurmath NR, Gaitonde VN. Performance studies on homogeneous charge compression ignition (HCCI) engine powered with alternative fuels. Renew Energy 2019;132:683–93.
[69] Paulauskiene T, Bucas M, Laukinaite A. Alternative fuels for marine applications: biomethanol-biodiesel-diesel blends. Fuel 2019;248:161–7.
[70] Fu J, Turn SQ. Characteristics and stability of biofuels used as drop-in replacement for NATO marine diesel. Fuel 2019;236:516–24.
[71] Dimitriou P, Tsujimura T, Suzuki Y. Adopting biodiesel as an indirect way to reduce the NO_x emission of a hydrogen fumigated dual-fuel engine. Fuel 2019;244:324–34.
[72] Bae C, Kim J. Alternative fuels for internal combustion engines. Fuel 2019;36: 3389–413.
[73] Dueso C, Muñoz M, Moreno F, Arroyo J, Gil-Lalaguna N, Bautista A, Gonzalo A, Sánchez JL. Adopting biodiesel as an indirect way to reduce the NOx emission of a hydrogen fumigated dual-fuel engine. Fuel 2018;234:276–85.
[74] Wang SM, Kuan DY, Du W, Zhao XB, Liu DH. Effect of organic acid treatment on microalgae lipid extraction. Chem J Chinese Universities 2015;36(3):499–504.
[75] Vyavhare K, Bagi S, Patel M, Aswath PB. Impact of diesel engine oil additives–soot interactions on physiochemical, oxidation, and wear characteristics of soot. Energy Fuel 2019;33(5):4515–30.
[76] Caplin A, Ghandehari M, Lim C, Glimcher P, Thurston G. Advancing environmental exposure assessment science to benefit society. Nat Commun 2019;10(1):1–11.
[77] Kontou A, Southby M, Spikes HA. Effect of steel hardness on soot wear. Wear 2017;390:236–45.
[78] Salehi FM, Khaemba DN, Morina A, Neville A. Corrosive–abrasive wear induced by soot in boundary lubrication regime. Tribol Lett 2016;63(2):19.
[79] Kontou A, Southby M, Morgan N, Spikes HA. Influence of dispersant and ZDDP on soot wear. Tribol Lett 2018;66(4):157.
[80] Xiong Y, Su P, Liu X, Yang H. Effects of diesel soot on the tribology behavior of organo-molybdenum friction modifier. Acta Petrol Sin 2018;34(5):125–32.
[81] Hu EZ, Hu XG, Liu TX, Fang L, Dearn KD, Xu HM. The role of soot particles in the tribological behavior of engine lubricating oils. Wear 2013;304(1):152–61.
[82] Guo MF, Cai ZB, Zhang ZC, Zhu MH. Characterization and lubrication performance of diesel soot nanoparticles as oil lubricant additives. RSC Adv 2015;123:101965–74.
[83] Ma LR, Zhang CH, Liu SH. Progress in experimental study of aqueous lubrication. Chin Sci Bull 2012;57(17):2062–9.
[84] Tang JZ, Chen SQ, Jia YL, Ma Y, Xie HM, Quan X, Ding Q. Carbon dots as an additive for improving performance in water-based lubricants for amorphous carbon (a-C) coatings. Carbon 2020;156:272–81.
[85] Huang GW, Yu QL, Ma ZF, Cai MR, Zhou F, Liu WM. Fluorinated candle soot as the lubricant additive of perfluoropolyether. Tribol Lett 2017;65(1):28.
[86] Gao CL, Abella L, Tian HR, Zhang X, Zhong YY, Tan YZ, Wu XZ. Double functionalization of a fullerene in drastic arc-discharge conditions: synthesis and formation mechanism of $C_{2v}(2)$-$C_{78}Cl_6(C_5Cl_6)$. Carbon 2018;129:286–92.
[87] Xu BS, Sun RP, Han PD, Li TB, Wang XM, Liu XG. Surface chemical modification of onion-like fullerenes. Chem J Chinese U 2006;27(8):1404–8.
[88] Jolypottuz L, Vacher B, Ohmae N, Martin JM, Epicier T. Anti-wear and friction reducing mechanisms of carbon nano-onions as lubricant additives. Tribol Lett 2008;31(1):69–80.
[89] Yao YL, Yi HH. Tribological properties of Fe/onion-like fullerenes as lubricant additives. J Chem Ind Eng 2011;62(1):281–6.
[90] Jia J, Zhang WK, Jiao C, Yang HY. Friction properties of carbon nano onions as lubricating oil additives. J Synth Cryst 2017;46(7):170–5.

[91] Tevet O, Vonhuth P, Popovitzbiro R, Rosentsveig R, Wagner HD. Friction mechanism of individual multilayered nanoparticles. Proc Natl Acad Sci USA 2011;108(50):19901–6.

[92] Deng X, Mammen L, Butt HJ, Vollmer D. Candle soot as a template for a transparent robust superamphiphobic coating. Science 2012;335(6064):67–70.

[93] Thomson MJ, Mitra T. A radical approach to soot formation. Science 2018; 361(6406):978–9.

[94] Mulay MR, Chauhan A, Patel S, Balakrishnan V, Halder A, Vaish R. Candle soot: journey from a pollutant to a functional material. Carbon 2019;144:684–712.

[95] Su Z, Zhou W, Zhang Y. New insight into the soot nanoparticles in a candle flame. Chem Commun 2011;47(16):4700–2.

[96] Singh J, Nandi T, Ghosh SK, Srivastava J, Tripathi SK, Prasad NE. Carbon nanoparticle synthesis, separation, characterization, and tribological property evaluation. Sep Sci Technol 2018;53(14):2314–26.

[97] Bucholz EW, Phillpot SR, Sinnott SB. Molecular dynamics investigation of the lubrication mechanism of carbon nano-onions. Comput Mater Sci 2012;54:91–6.

[98] Li C, Wu B, Liu TX, Liu YM, Hu XG. Physical-chemical and aggregation properties of biodiesel soot particles prepared at normal temperature and atmospheric pressure. Trans Chin Soc Agric Eng 2017;33(13):90–6.

[99] Mutyala KC, Wu YA, Erdemir A, Sumant AV. Graphene-MoS_2 ensembles to reduce friction and wear in DLC-Steel contacts. Carbon 2019;146:524–7.

[100] Li CC, Huang CL. Preparation of clear colloidal solutions of detonation nanodiamond in organic solvents. Colloids Surf A 2010;353(1):52–6.

[101] Li XQ, Sun CY, Zhao Q, Wu ZS, Xu DK, Zhong WY. Preparation of clear colloidal solutions of detonation nanodiamond in organic solvents. Chem J Chinese Universities 2013;34(4):763–70.

[102] Seymour BT, Wright RA, Parrott AC, Gao HY, Martini A, Qu J, Dai S, Zhao B. Poly(alkyl methacrylate) brush-grafted silica nanoparticles as oil lubricant additives: effects of alkyl pendant groups on oil dispersibility, stability, and lubrication property. ACS Appl Mater Interfaces 2017;9(29):25038–48.

[103] Chen Q, Wang X, Wang ZT, Liu Y, You TZ. Preparation of water-soluble nanographite and its application in water-based cutting fluid. Nanoscale Res Lett 2013;8(1):52.

[104] Li K, Lin CX, Li W, Zhou ZR. Frictional characteristics of esophageal tissue under different lubricant media. Tribology 2018;38(5):32–8.

[105] Xie C, Hong GH, Yang WQ, Wang JK, Zhao LN. Antibacterial superhydrophobic-oleophobic coating fabricated by candle soot. Chem J Chinese U 2019;40(2):379–84.

[106] Xie H, Jiang B, Dai JH, Peng C, Li CX, Li Q, Pan FS. Tribological behaviors of graphene and graphene oxide as water-based lubricant additives for magnesium alloy/steel contacts. Materials 2018;11(2):206.

[107] He AS, Huang SQ, Yun JH, Jiang ZY, Stokes JR, Jiao SH, Wang LZ, Huang H. Tribological characteristics of aqueous graphene oxide, graphitic carbon nitride, and their mixed suspensions. Tribol Lett 2018;66(1):42.

[108] Meng Y, Su FH, Chen YZ. Au/Graphene oxide nanocomposite synthesized in supercritical CO_2 fluid as energy efficient lubricant additive. Tribol Lett 2017;65(3):104.

Index

Note: Page numbers followed by *f* indicate figures and *t* indicate tables.

FSC
www.fsc.org
MIX
Papier | Fördert gute Waldnutzung
FSC® C083411

Kolibri 360 GmbH
Ferdinand-Jühlke-Straße 7
99095 Erfurt, Deutschland
produktsicherheit@kolibri360.de